HIGH DYNAMIC RANGE VIDEO

Synthesis Lectures on Computer Graphics and Animation

Editor
Brian A. Barsky, *University of California, Berkeley*

High Dynamic Range Video
Karol Myszkowski, Rafał Mantiuk, and Grzegorz Krawczyk
2008

GPU-Based Techniques For Global Illumination Effects
László Szirmay-Kalos, László Szécsi and Mateu Sbert
2008

High Dynamic Range Imaging Reconstruction
Asla Sa, Paulo Carvalho, and Luiz Velho IMPA, Brazil
2007

High Fidelity Haptic Rendering
Miguel A. Otaduy, Ming C. Lin
2006

A Blossoming Development of Splines
Stephen Mann
2006

High Dynamic Range Video

Karol Myszkowski, Rafał Mantiuk, and Grzegorz Krawczyk

ISBN: 978-3-031-79527-5   paperback
ISBN: 978-3-031-79528-2   ebook

DOI: 10.1007/978-3-031-79528-2

A Publication in the Springer series

SYNTHESIS LECTURES ON COMPUTER GRAPHICS AND ANIMATION #5

Lecture #5

Series Editor: Brian A. Barsky, University of California, Berkeley

Library of Congress Cataloging-in-Publication Data

Series ISSN: 1933-8996   print
Series ISSN: 1933-9003   electronic

HIGH DYNAMIC RANGE VIDEO

Karol Myszkowski, Rafał Mantiuk, and Grzegorz Krawczyk

SYNTHESIS LECTURES ON COMPUTER GRAPHICS AND ANIMATION #5

ABSTRACT

As new displays and cameras offer enhanced color capabilities, there is a need to extend the precision of digital content. High Dynamic Range (HDR) imaging encodes images and video with higher than normal 8 bit-per-color-channel precision, enabling representation of the complete color gamut and the full visible range of luminance. However, to realize transition from the traditional to HDR imaging, it is necessary to develop imaging algorithms that work with the high-precision data. To make such algorithms effective and feasible in practice, it is necessary to take advantage of the limitations of the human visual system by aligning the data shortcomings to those of the human eye, thus limiting storage and processing precision. Therefore, human visual perception is the key component of the solutions we discuss in this book.

This book presents a complete pipeline for HDR image and video processing from acquisition, through compression and quality evaluation, to display. At the HDR image and video acquisition stage specialized HDR sensors or multi-exposure techniques suitable for traditional cameras are discussed. Then, we present a practical solution for pixel values calibration in terms of photometric or radiometric quantities, which are required in some technically oriented applications. Also, we cover the problem of efficient image and video compression and encoding either for storage or transmission purposes, including the aspect of backward compatibility with existing formats. Finally, we review existing HDR display technologies and the associated problems of image contrast and brightness adjustment. For this purpose tone mapping is employed to accommodate HDR content to LDR devices. Conversely, the so-called inverse tone mapping is required to upgrade LDR content for displaying on HDR devices. We overview HDR-enabled image and video quality metrics, which are needed to verify algorithms at all stages of the pipeline. Additionally, we cover successful examples of the HDR technology applications, in particular, in computer graphics and computer vision.

The goal of this book is to present all discussed components of the HDR pipeline with the main focus on video. For some pipeline stages HDR video solutions are either not well established or do not exist at all, in which case we describe techniques for single HDR images. In such cases we attempt to select the techniques, which can be extended into temporal domain. Whenever needed, relevant background information on human perception is given, which enables better understanding of the design choices behind the discussed algorithms and HDR equipment.

KEYWORDS

High Dynamic Range Imaging (HDRI), Tone Mapping, Tone Reproduction, Inverse Tone Mapping, HDR Display Devices, HDR Camera Sensors, Multi-exposure HDR Image Capture, HDR Image Acquisition, HDR Video Acquisition, Radiometric (Photometric) Camera Calibration, HDR Image and Video Compression, Scalable Bit-depth Coding, HDR Image File Formats, Image Quality Metrics, Image-Based Lighting, 3D Object Appearance Acquisition.

Contents

Acknowledgements xi

1. Introduction 1
   1.1 Low versus High Dynamic Range Imaging 1
   1.2 Device- and Scene-Referred Image Representations 3
   1.3 HDR Revolution 4
   1.4 Organization of the Book 6
       1.4.1 Why HDR Video? 7
       1.4.2 Chapter Overview 8

2. Representation of an HDR Image 9
   2.1 Light 9
   2.2 Color 11
   2.3 Dynamic Range 15

3. HDR Image and Video Acquisition 17
   3.1 Capture Techniques Capable of HDR 17
       3.1.1 Temporal Exposure Change 17
       3.1.2 Spatial Exposure Change 19
       3.1.3 Multiple Sensors with Beam Splitters 20
       3.1.4 Solid-State Sensors 20
   3.2 Photometric Calibration of HDR Cameras 21
       3.2.1 Camera Response to Light 22
       3.2.2 Mathematical Framework for Response Estimation 22
       3.2.3 Procedure for Photometric Calibration 25
       3.2.4 Example Calibration of HDR Video Cameras 27
       3.2.5 Quality of Luminance Measurement 30
       3.2.6 Alternative Response Estimation Methods 31
       3.2.7 Discussion 32

4. HDR Image Quality 34
   4.1 Visual Metric Classification 35
   4.2 A Visual Difference Predictor for HDR Images 37
       4.2.1 Implementation 40

5. HDR Image, Video, and Texture Compression 41

5.1 HDR Pixel Formats and Color Spaces 42

    5.1.1 Minifloat: 16-Bit Floating Point Numbers 43

    5.1.2 RGBE: Common Exponent 43

    5.1.3 LogLuv: Logarithmic Encoding 44

    5.1.4 RGB Scale: Low-Complexity RGBE Coding 45

    5.1.5 LogYuv: Low-Complexity LogLuv 46

    5.1.6 JND Steps: Perceptually Uniform Encoding 46

5.2 High Fidelity Image Formats 50

    5.2.1 Radiance's HDR Format 51

    5.2.2 OpenEXR 51

5.3 High Fidelity Video Formats 52

    5.3.1 Digital Motion Picture Production 52

    5.3.2 Digital Cinema 53

    5.3.3 MPEG for High-Quality Content 53

    5.3.4 HDR Extension of MPEG-4 54

5.4 Backward-Compatible Compression 55

    5.4.1 JPEG HDR 55

    5.4.2 Wavelet Compander 56

    5.4.3 Backward-Compatible HDR MPEG 57

    5.4.4 Scalable High Dynamic Range Video Coding from the JVT 61

5.5 High Dynamic Range Texture Compression 62

5.6 Conclusions 64

6. Tone Reproduction 67

6.1 Tone-Mapping Operators 67

    6.1.1 Luminance Domain Operators 68

    6.1.2 Local Adaptation 69

    6.1.3 Prevention of Halo Artifacts 71

    6.1.4 Segmentation-Based Operators 72

    6.1.5 Contrast Domain Operators 73

6.2 Tone-Mapping Studies with Human Subjects 76

6.3 Objective Evaluation of Tone Mapping 79

    6.3.1 Contrast Distortion in Tone Mapping 79

    6.3.2 Analysis of Tone-Mapping Algorithms 80

6.4 Temporal Aspects of Tone Reproduction 83

6.5 Conclusions 86

7.   **HDR Display Devices** 89
  7.1   HDR Display Requirements 89
  7.2   Dual-Modulation Displays 91
  7.3   Laser Projection Systems 95
  7.4   Conclusions 96

8.   **LDR2HDR: Recovering Dynamic Range in Legacy Content** 99
  8.1   Bit-Depth Expansion and Decontouring Techniques 100
  8.2   Reversing Tone-Mapping Curve 104
  8.3   Single Image-Based Camera Response Approximation 107
  8.4   Recovering Clipped Pixels 110
  8.5   Handling Video on-the-Fly 111
  8.6   Exploiting Image Capturing Artifacts for Upgrading Dynamic Range 113
  8.7   Conclusions 114

9.   **HDRI in Computer Graphics** 115
  9.1   Computer Graphics as the Source of HDR Images and Video 115
  9.2   HDR Images and Video as the Input Data for Computer Graphics 119
     9.2.1   HDR Video-Based Lighting 119
     9.2.2   HDR Imaging in Reflectance Measurements 133
  9.3   Conclusions 140

10. **Software** 143
  10.1 pfstools 143
  10.2 pfscalibration 144
  10.3 pfstmo 144
  10.4 HDR Visible Differences Predictor 144

   Bibliography 145

   Author Biography 157

Acknowledgements

A project like this book could not have been completed without the support from many colleagues, friends, and family. We would like to thank Hans-Peter Seidel for his encouragement and support for our research projects in the novel field of High Dynamic Range (HDR) imaging that we conducted at the Max-Planck Institute. Many of these projects would not have been possible without the help and collaboration of our colleagues at the Max-Planck Institute: Tunç Ozan Aydin, Michael Goesele, Akiko Yoshida, Kaleigh Smith, Robert Herzog, Volker Blanz, Tom Annen, Vlastimil Havran, Mirosław Smyk, Tina Scherbaum, Thorsten Grosch, Tobias Ritschel, Matthias Ihrke, Gernot Ziegler, Radosław Mantiuk, Andrei Lintu, Hendrik Lensch, Martin Fuchs, Art Tevs, Ivo Ihrke, Christian Theobalt, Christian Fuchs, Anna Tomaszewska, and Dorota Zdrojewska.

Our interactions with several exceptional colleagues had significant impact on the development of this book. We have learned a great deal about HDR issues through insightful discussions with Scott Daly, Wolfgang Heidrich, Greg Ward, Sumant Pattanaik, Alan Chalmers, Erik Reinhard, Helge Seetzen, Bernd Höfflinger, John McCann, and Alessandro Rizzi. In particular, we are grateful to Helge Seetzen for making available to us the whole family of prototype HDR displays developed by BightSide Tech. We would like also to thank Bernd Höfflinger, Volker Gengenbach, and Daniel Brosch for providing us with HDR cameras developed at IMS-CHIPS, which let us generate the first HDR video content for our video compression and dynamic scene re-lighting projects. Also, we would like to express our deep appreciation to anonymous reviewers for their insightful comments, which greatly improved this book. We thank the authors of images who kindly agreed that we use their work in this book.

We would like to thank the people from Morgan & Claypool Publishing. In particular, we are grateful to Michael Morgan for coming to us with the idea of writing this book and for his tireless patience in the course of making this happen. We thank Samir Roy for the great job he has done producing this book.

Last, but not least, we would like to thank our families and close friends, who supported us throughout writing this book.

CHAPTER 1

Introduction

1.1 LOW VERSUS HIGH DYNAMIC RANGE IMAGING

The majority of existing digital imagery and video material capture only a fraction of the visual information that is visible to the human eye and are not of sufficient quality for reproduction by the future generation of display devices. The limiting factor is not the resolution, since most consumer level digital cameras can take images of higher number of pixels than most of displays can offer. The problem is the limited color gamut and even more limited dynamic range (contrast) captured by cameras and stored by the majority of image and video formats. To emphasize these limitations of traditional imaging technology, it is often called *low-dynamic range* or simply *LDR*.

For instance, each pixel value in the JPEG image encoding is represented using three 8-bit integer numbers (0–255) using the YC_rC_b color space. This color space is able to store only a small part of visible color gamut (although containing the colors most often encountered in the real world), as illustrated in Fig. 1.1-left, and an even smaller part of the luminance range that can be perceived by our eyes, as illustrated in Fig. 1.1-right. The reason for this is that the JPEG format was designed to store as much information as can be displayed on the majority of displays, which were at that time cathode ray tube (CRT) monitors or TV sets. This assumption is no longer valid, as the new generations of LCD and Plasma displays can depict a much broader color gamut and dynamic range than their CRT ancestors. Every new generation of displays offers better color reproduction and requires higher precision of image and video content. The traditional low contrast range and limited color gamut imaging (LDR imaging), which is confined to three 8-bit integer color channels, cannot offer the precision that is needed for the upcoming developments in image capture, processing, storage, and display technologies.

High dynamic range imaging (HDRI) overcomes the limitation of traditional imaging by performing operations on color data with much higher precision. Pixel colors are specified in HDR images as a triple of floating point values (usually 32-bit per color channel), providing accuracy that exceeds the capabilities of the human visual system under any viewing conditions.

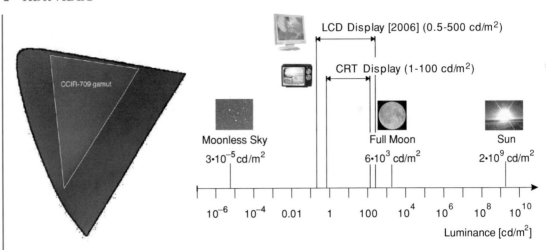

FIGURE 1.1: Left: the standard color gamut frequently used in traditional imaging (CCIR-705), compared to the full visible color gamut. Right: real-world luminance values compared with the range of luminance that can be displayed on CRT and LDR monitors. Most digital content is stored in a format that at most preserves the dynamic range of typical displays.

By its inherent colorimetric precision, HDRI can represent all colors found in real world that can be perceived by the human eye.

HDRI does not only provide higher precision, but also enables the synthesis, storage and visualization of a range of perceptual cues that are not achievable with traditional imaging. Most of the LDR imaging standards and color spaces have been developed to match the needs of office or display illumination conditions. When viewing such scenes or images under such conditions, our visual system operates in a mixture of day-light and dim-light vision state, so-called the mesopic vision. When viewing outdoor scenes, we use day-light perception of colors, so-called the photopic vision. This distinction is important for digital imaging as both types of vision show different performance and result in different perception of colors. HDRI can represent images of the luminance range fully covering both the photopic and the mesopic vision, thus making distinction between them possible. One of the differences between mesopic and photopic visions is the impression of colorfulness. We tend to regard objects more colorful when they are brightly illuminated, which is the phenomenon that is called Hunt's effect. To render enhanced colorfulness properly, digital images must preserve information about the actual level of luminance of the original scene, which is not possible in the case of traditional imaging.

Real-world scenes are not only brighter and more colorful than their digital reproductions, but also contain much higher contrast, both local between neighboring objects, and global between distant objects. The eye has evolved to cope with such high contrast and its presence in a scene evokes important perceptual cues. Traditional imaging, unlike HDRI, is not able to represent such high-contrast scenes. Similarly, traditional images can hardly represent common visual phenomena, such as self-luminous surfaces (sun, shining lamps) and bright specular highlights. They also do not contain enough information to reproduce visual glare (brightening of the areas surrounding shining objects) and a short-time dazzle due to sudden increase of the brightness of a scene (e.g., when exposed to the sunlight after staying indoors). To faithfully represent, store and then reproduce all these effects, the original scene must be stored and treated using high fidelity HDR techniques.

1.2   DEVICE- AND SCENE-REFERRED IMAGE REPRESENTATIONS

To accommodate all discussed requirements imposed on HDRI, a common format of data is required to enable their efficient transfer and processing on the way from HDR acquisition to HDR display devices. Here again fundamental differences between image formats used in traditional imaging and HDRI arise, which we address in this section.

Commonly used LDR image formats (JPEG, PNG, TIFF, etc.) contain data that are tailored to particular display devices: cameras, CRT or LCD displays. For example, two JPEG images shown using two different LCD displays may be significantly different due to dissimilar image processing, color filters, gamma correction, and so on. Obviously, such representation of images vaguely relates to the actual photometric properties of the scene it depicts, but it is dependent on a display device. Therefore, those formats can be considered as *device-referred* (also known as *output-referred*), since they are tightly coupled with the capabilities and characteristic of a particular imaging device.

ICC color profiles can be used to convert visual data from one device-referred format to another. Such profiles define the colorimetric properties of a device for which the image is intended for. Problems arise if the two devices have different color gamuts or dynamic ranges, in which case a conversion from one format to another usually involves the loss of some visual information. The algorithms for the best reproduction of LDR images on the output media of different color gamuts have been thoroughly studied [1] and CIE technical committee (CIE Division 8: TC8-03) have been started to choose the best algorithm. However, as for now, the committee has not been able to select a single algorithm that would give reliable results in all cases. The problem is even more difficult when an image captured with an HDR camera is converted to the color space of a low-dynamic range monitor (see a multitude of tone reproduction algorithms discussed in Chapter 6). Obviously, the ICC profiles cannot be easily used to facilitate interchange of data between LDR and HDR devices.

Scene-referred representation of images offers a much simpler solution to this problem. The scene-referred image encodes the actual photometric characteristic of a scene it depicts. Conversion from such common representation, which directly corresponds to physical luminance or spectral radiance values, to a format suitable for a particular device is the responsibility of that device. This should guarantee the best possible rendering of the HDR content, since only the device has all the information related to its limitations and sometimes also viewing conditions (e.g., ambient illumination), which is necessary to render the content properly. HDR file formats are examples of scene-referred encoding, as they usually represent either luminance or spectral radiance, rather than gamma corrected and ready to display "pixel values."

The problem of accuracy of scene-referred image representation arises, for example the magnitude of quantization error and its distribution for various luminance levels in the depicted scene. For display-referred image formats the problem of pixel accuracy is easy to formulate in terms of the reproduction capabilities of target display devices. For scene-referred image representations, the accuracy should not be tailored to any particular imaging technology and, if efficiency of storing data is required, should be limited only by the capabilities of the human visual system.

To summarize, the difference between HDRI and traditional LDR imaging is that HDRI always operates on device-independent and high-precision data, so that the quality of the content is reduced only at the display stage, and only if a device cannot faithfully reproduce the content. This is contrary to traditional LDR imaging, where the content is usually profiled for particular device and thus stripped from useful information as early as at the acquisition stage or latest at the storage stage. Figure 1.2 summarizes these basic conceptual differences between LDR and HDR imaging.

1.3   HDR REVOLUTION

HDRI has recently gained momentum and is affecting almost all fields of digital imaging. One of the breakthroughs responsible for this burst of interest in HDRI was the development of an HDR display, which proved that the visualization of color and the luminance range close to real-world scenes is possible [2]. One of the first to adopt HDRI was video game developers together with graphics card vendors. Today most of the state-of-the art video game engines perform rendering using HDR precision to deliver more believable and appealing virtual reality imagery. Computer-generated imagery used in special effect production uses HDR techniques to achieve the best match between synthetic and realistic objects. High-end cinematographic cameras, both analog and digital, already provide significantly higher dynamic range than most of the displays today. This dynamic range can be retained after digitalization only if a form of HDR representation is used. HDRI is also a strong trend in digital photography, mostly due to the multi-exposure techniques that allow an HDR image to be made using a consumer level digital camera. HDR cameras that can directly capture the higher dynamic range are available,

Standard (Low) Dynamic Range		**High Dynamic Range**
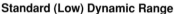		
50 dB	*Camera Dynamic Range*	120 dB
1:200	*Display Contrast*	1:15,000
8-bit or 16-bit	*Quantization*	floating point or variable
display-referred	*Image Representation*	scene-referred
display-limited	*Fidelity*	as good as the eye can see

FIGURE 1.2: The advantages of HDR compared to LDR from the applications point of view. The quality of the LDR image has been reduced on purpose to illustrate a potential difference between the HDR and LDR visual contents as seen on an HDR display. The given numbers serve as an example and are not meant to be a precise reference.

for example *SheroCamHDR* from *SpheronVR*, *HDRC* from *IMS Chips*, *Origin*®from *Dalsa* or *Viper FilmStream* from *Thomson*. Also, major display vendors experiment with local dimming technology and LED-based backlight devices, which significantly enhances the dynamic range of offered by them LCD displays. To catch up with the HDR trend, many software vendors announce their support of the HDRI, taking *Adobe*® *Photoshop*® *CS3* and *Corel*® *Paint Shop Pro*® *Photo X2* as examples. Also, commercial packages supporting multi-exposure blending and tone reproduction such as *Photomatix* or *FDRTools* targeted mostly for photographers become available.

Besides its significant impact on existing imaging technologies that we can observe today, HDRI has the potential to radically change the methods by which imaging data are processed, displayed, and stored in several fields of science. Computer vision algorithms can greatly benefit from the increased precision of HDR images, which do not have over- or under-exposed regions and which are often the cause of the algorithms failure. Medical imaging has already developed image formats (e.g. the DICOM format) that partly cope with the shortcomings of traditional images; however, they are supported only by specialized hardware and software. HDRI gives the sufficient precision for medical imaging and therefore its capture, processing, and rendering techniques can be used also in this field. HDR techniques can also find applications in astronomical imaging, remote sensing, industrial design, and scientific visualization.

All these exciting developments in HDRI as well as huge potential of this technology in multiple applications suggest that imaging is on the verge of HDR revolution. This revolution will have a profound impact on devices that are used for image capture and display, as well as on image and video formats that are used to store and broadcast visual content. Obviously, during the transition time some elements of imaging pipeline may still rely on traditional LDR technology. This will require backward compatibility of HDR formats to enable their use on LDR output devices such as printers, displays, and projectors. For some of such devices, the format extensions to HDR should be transparent, and standard *display-referred* content should be directly accessible. However, more advanced LDR devices may take advantage of HDR information by adjusting *scene-referred* content to their technical capabilities through customized tone reproduction. Finally, the legacy images and video should be upgraded when displayed on HDR devices, so that the best possible image quality is achieved (the so-called inverse tone mapping). In this book, we address all these important issues by focusing mostly on the state-of-the-art techniques. An interesting account of historical developments on dynamic range expansion in the art, traditional photography, and electronic imaging has been recently presented by one of the pioneers in HDRI John McCann [3].

1.4   ORGANIZATION OF THE BOOK

The book presents a complete pipeline for HDR image and video processing from acquisition, through compression and quality evaluation, to display (refer to Fig. 1.3). At the first stage digital images are acquired, either with cameras or computer rendering methods. In the former case, pixel values calibration in terms of photometric or radiometric quantities may be required in some technically oriented applications. At the second stage, digital content is efficiently compressed and encoded either for storage or transmission purposes. Here backward compatibility with existing formats is an important issue. Finally, digital video or images are displayed on display devices. Tone mapping is required to accommodate HDR content to LDR devices, and conversely LDR content upgrading (the so-called inverse tone mapping) is necessary for displaying on HDR devices. Apart from considering technical capabilities of display devices, the viewing conditions such as ambient lighting and amount of light reflected by the display play an important role for proper determination of tone-mapping parameters. Quality metrics are employed to verify algorithms at all stages of the pipeline.

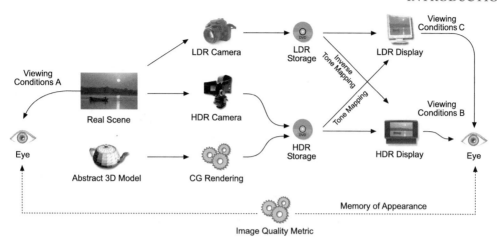

FIGURE 1.3: Imaging pipeline and available HDR technologies.

Additionally, the book includes successful examples of the use of HDR technology in research setups and industrial applications involving computer graphics. Whenever needed short background information on human perception is given, which enables better understanding of the design choices behind the discussed algorithms and HDR equipment.

The goal of this book is to present all discussed components of the HDR pipeline with the main focus on HDR video. For some pipeline stages HDR video solutions are not well established or do not exist at all, in which case we describe techniques for single HDR images. In such cases, we attempt to select the techniques, which can be extended into temporal domain.

1.4.1 Why HDR Video?

Our focus in this book on HDR video stems from the fact that while HDR images are visually compelling and relatively common (over 125 000 photographs tagged as HDR is available on Flickr), the key applications that will drive further HDRI development in coming years require some form of HDR video or uncompressed temporal image sequences. It can be envisioned that the entertainment industry with computer games, digital cinema, and special effects will be such an important driving force. In games due to HDR-enabled (floating point) graphics pipelines HDR image sequences can be readily generated as an output from modern GPU cards. In the near future, games will use more often HDR video of real-world scenes for virtual scenes relighting or as realistic video textures. In digital cinema applications, the lack of desirable contrast and luminance range are the main current drawbacks, whose prompt improvement can be expected in the quest for a better visual quality than it is possible with traditional film projectors. In terms of HDR content for digital cinema, this does not look like a real problem.

Modern movies have often been shot with cameras featuring a higher dynamic range, and legacy movies can be upgraded even if manual intervention would be required for some frames (as this happened in the past with black&white films' upgrade to color). Also, special effects, especially those in which real and synthetic footage are seamlessly mixed, require both HDR shooting and rendering. HDR video is also required in all applications in which capturing temporal aspects of changes in the scene is required with high accuracy. This is in particular important in monitoring of some industrial processes such as welding, predictive driver assistance systems in automotive industry, surveillance systems, to name just a few possible applications. HDR video can be also considered to speed up the image acquisition in all applications, in which a large number of static HDR images are required, as for example in image-based techniques in computer graphics. Finally, with the spread of TV sets featuring enhanced dynamic range, broadcasting of HDR video signal will be important, which may take long time before it actually happens due to standardization issues. For this particular application, enhancing current LDR video signal to HDR by intelligent TV sets seems to be a viable solution in the nearest future.

1.4.2 Chapter Overview

The book is organized as follows: Chapter 2 gives background information on the digital representation of images and the photometric and colorimetric description of light and color. Chapter 3 reviews the HDR image and video capture techniques and describes the procedure of their photometric calibration, so that the pixel values are directly expressed in luminance units. Chapter 4 presents a perception-based image quality metric, which enables the prediction of differences between a pair of HDR images. Such metrics are important to judge the quality of HDR content for example as the result of lossy compression. Chapter 5 discusses the issues of HDR image and video compression. At first HDR pixel format and color spaces are reviewed and then existing formats of HDR image and video encoding are presented. Special attention is paid to backward-compatible compression schemes. Chapter 6 presents a synthetic overview of state-of-the-art tone-mapping operators and discusses the problem of their evaluation using subjective methods with human subjects and objective computational models. Also, temporal aspects of tone reproduction are investigated. Chapter 7 briefly surveys HDR display and projection technologies that appeared in recent years. The problem of upgrading legacy images and video (inverse tone mapping), so that they can be displayed on HDR devices with the best visual quality, is discussed in Chapter 8. Chapter 9 surveys cross-correlations between developments in computer graphics and HDRI. At first, computer graphics rendering as a rich source of high quality HDR content is presented. Then, HDR images and video captured in the real world as the input data for image-based rendering and modeling are discussed. Finally, Chapter 10 demonstrates software packages for processing of HDR images and video that have been made available by the authors of this book as open-source projects.

CHAPTER 2

Representation of an HDR Image

This chapter explains several physical and perceptual quantities important for digital imaging, such as radiance, luminance, luminance factor, and color. It does not give a complete or exhaustive introduction to radiometry, photometry or colorimetry, since these are described in full extent elsewhere [4,5,6]. The focus of this chapter is on the concepts that are confusing or vary in terminology between disciplines, and also those that are used in the following chapters.

2.1 LIGHT

The physical measure of light that is the most appropriate for imaging systems is either *luminance* (used in photometry) or *spectral radiance* (used in radiometry). This is because both measures stay constant regardless of the distance from a light source to a sensor (assuming no influence of the medium in which the light travels). The sensor can be either camera's CCD chip or a photoreceptor in the eye. The quantities measured by photoreceptors or digital sensors are related to either of these measures.

Spectral radiance is a radiometric measure, defined by

$$L(\lambda) = \frac{d^2 \Phi(\lambda)}{d\omega \cdot dA \cdot \cos \theta},$$

(2.1)

where $L(\lambda)$ is spectral radiance for the wavelength λ, Φ is the radiant flux flowing through a surface per unit time, ω is the solid angle, θ is the angle between the rays and the surface, and A

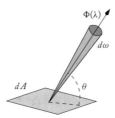

FIGURE 2.1: Spectral radiance. Spectral radiance is a differential measure, defined for infinitely small area dA, infinitely small solid angle $d\omega$, radiant flux Φ and an angle between the rays and the surface θ.

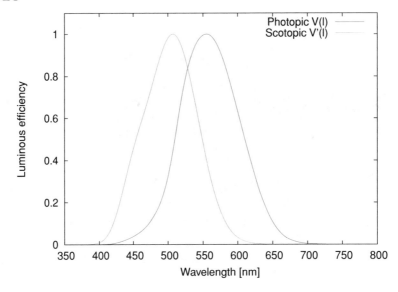

FIGURE 2.2: CIE spectral luminous efficiency curve for photopic (day light) and scotopic (night) vision. Data downloaded from http://www.cvrl.org/.

is the area of the surface, as illustrated in Fig. 2.1. Although spectral radiance is commonly used in computer graphics, images are better defined with photometric units of *luminance*. *Luminance* is spectral radiance integrated over the range of visible wavelengths with the weighting function $V(\lambda)$:

$$Y = \int_{380\,nm}^{770\,nm} L(\lambda)V(\lambda)d\lambda. \tag{2.2}$$

The function $V(\lambda)$, which is called the *spectral luminous efficiency curve* [7], gives more weight to the wavelengths, to which the human visual system (HVS) is more sensitive. This way luminance is related (though non-nonlinearly) to our perception of brightness. The function V for the daylight vision (photopic) and night vision (scotopic) is plotted in Fig. 2.2. The temporal aspects of daylight and night vision will be discussed in more detail in Section 6.4. Luminance, Y, is usually given in cd/m^2 or equivalent *nit* units.

Since the most common multi-exposure technique for acquiring HDR images (refer to Section 3.1.1) cannot assess the absolute luminance level but only a relative luminance values, most HDR images do not contain luminance values but rather the values of *luminance factor*. Such luminance factor must be multiplied by a constant number, which depends on a camera and lens, to get actual luminance. Such constant number can be easily found if we can measure the luminance of a photographed surface (refer to Section 3.2).

2.2   COLOR

Colors are perceptual phenomena rather than physical. Although we can precisely describe colors using physical units of spectral radiance, such description does not give immediate answer whether the described color is green or red. *Colorimetry* is the field that numerically characterizes colors and provides a link between the human color perception and the physical description of the light. This section introduces the most fundamental aspects of colorimetry and introduces color spaces, which will be used in later chapters. More detailed introduction to colorimetry can be found in [8] and [6], while two handbooks, [5] and [4], are more exhaustive source of information.

The human color perception is determined by three types of cones: L, M, and S, and their sensitivity to wavelengths. The light in the visible spectrum is in fact multi-dimensional variable, where each dimension is associated with particular wavelength. However, the visible color is a projection of this multi-dimensional variable to three primaries, corresponding to three types of cones. Such projection is mathematically described as a product of the spectral power distribution, $\phi(\lambda)$, and the spectral response of the type of cones, $C_L(\lambda)$, $C_M(\lambda)$ and $C_S(\lambda)$:

$$R = \int_\lambda \phi(\lambda)C_L(\lambda)d\lambda \tag{2.3}$$

$$G = \int_\lambda \phi(\lambda)C_M(\lambda)d\lambda \tag{2.4}$$

$$B = \int_\lambda \phi(\lambda)C_S(\lambda)d\lambda. \tag{2.5}$$

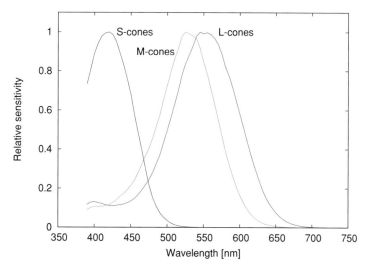

FIGURE 2.3: Cone photocurrent spectral responsivities. Redrawn from [9].

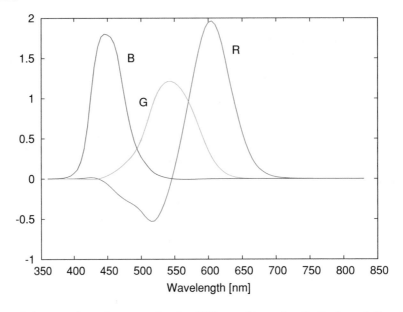

FIGURE 2.4: Color matching functions for the CIE matching stimuli *R*, *G*, and *B* and 2° standard observer. Data downloaded from http://www.cvrl.org/.

The spectral responsivities of cones are shown in Fig. 2.3.

As the result of three-dimensional encoding of color in the HVS, the number of distinguishable colors is limited. Also, two stimuli of different spectral power distributions can be seen as having the same color if only their *R*, *G*, and *B* projections match. The latter property of the HVS is called *metamerism*.

To uniquely describe visible color gamut, CIE standardized in 1931 a set of primaries for the standard colorimetric observer. Since the cone spectral responsivities were not known at that time, the primaries were based on color matching experiment, in which monochromatic stimuli of particular wavelength was matched with a mixture of the three monochromatic primaries (435.6 nm, 546.1 nm, and 700 nm). The values of color-matching mixture of primaries for each wavelength gave the *R*, *G*, and *B* primaries shown in Fig. 2.4. The drawback of this procedure was that it resulted in negative value of *R* primary. The negative part represents out of gamut colors, which are too saturated to be within visible or physically feasible range. To bring those colors into the valid gamut, the colors must be desaturated by adding monochromatic light. Since adding monochromatic light results in increasing the values of all *R*, *G*, and *B* components, there is a certain amount of the added light that would make all components positive.

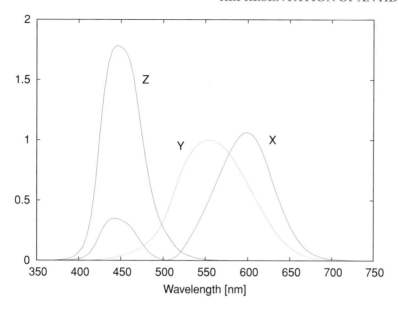

FIGURE 2.5: Color matching functions for the CIE matching stimuli X, Y, and Z and $2°$ standard observer. Data downloaded from `http://www.cvrl.org/`.

To avoid negative primaries and to connect colorimetric description of the light with photometric measure of luminance (see the previous section), CIE introduced XYZ primaries in 1931. The primaries, shown in Fig. 2.5, were designed so that primary Y represents luminance and its color matching function corresponds to the luminous efficiency function (see Fig. 2.2). Although the standard has been established over 70 years ago, it is still commonly used today, especially as a reference in color conversion formulas.

For a convenient two-dimensional representation of the color, chromaticity coordinates are often used:

$$x = \frac{X}{X+Y+Z} \tag{2.6}$$

$$y = \frac{Y}{X+Y+Z}. \tag{2.7}$$

Such coordinates must be accompanied by the corresponding luminance value, Y, to fully describe the color.

The visible differences between colors are not well described by chromaticity coordinates x and y. For better representation of perceptual color differences, CIE defined uniform chromaticity scales (UCS) in 1976, which are known as CIE 1976 Uniform Chromacity Scales:

$$u' = \frac{4X}{X + 15Y + 3Z} \tag{2.8}$$

$$v' = \frac{9Y}{X + 15Y + 3Z}. \tag{2.9}$$

Note that u', v' chromaticity space only approximates perceptual uniformity and a unit Cartesian distance may vary from 1 JND[1] to 4 JND units.

The Uniform Chromacity Scales do not incorporate luminance level in their description of color. This is a significant limitation, as color difference can strongly depend on the actual luminance level. Uniform color spaces have been introduced to address this problem. The first color space, CIE 1976 $L^*a^*b^*$, is defined by

$$L^* = 116(Y/Y_n)^{1/3} - 16 \tag{2.10}$$
$$a^* = 500\left[(X/X_n)^{1/3} - (Y/Y_n)^{1/3}\right] \tag{2.11}$$
$$b^* = 200\left[(Y/Y_n)^{1/3} - (Z/Z_n)^{1/3}\right] \tag{2.12}$$

and the second color space, CIE 1976 $L^*u^*v^*$, by

$$L^* = 116(Y/Y_n)^{1/3} - 16 \tag{2.13}$$
$$u^* = 13L^*(u' - u'_n) \tag{2.14}$$
$$v^* = 13L^*(v' - v'_n). \tag{2.15}$$

The coordinates with the n subscript denote the color of the *reference white*, which is the color that appears white in the scene. For color print, this is usually the color of a white paper under given illumination. Both color spaces have been standardized as the studies did not show that the one is definitely better over another and each one has its advantages.

Both CIE 1976 $L^*a^*b^*$ and CIE 1976 $L^*u^*v^*$ color spaces have been designed for low dynamic range color range, available on print or typical CRT displays and cannot be used for HDR images. In Section 5.1, we address this problem in more detail and in particular we derive an (approximately) perceptually uniform color space for HDR pixel values.

The uniform color spaces are the simplest incarnations of color appearance models. Color appearance models try to predict not only the colorimetric properties of the light, but also its appearance under given viewing conditions (background color, surround ambient light, color adaptation, etc.). CIECAM02 [10] is an example of such a model that has been standardized by CIE. The discussion of color appearance models would go beyond scope of this book; therefore, reader should refer to [4] and [8] for more information.

---

[1]JND—Just Noticeable Difference is usually defined as a measure of contrast at which a subject has 75% chance of correctly detecting visual difference in a stimulus.

TABLE 2.1: Measures of dynamic range and their context of application. The example column illustrates the same dynamic range expressed in different units.

NAME	FORMULA	EXAMPLE	CONTEXT
contrast ratio	$CR = 1 : (Y_{peak}/Y_{noise})$	1:500	displays
log exposure range	$D = \log_{10}(Y_{peak}) - \log_{10}(Y_{noise})$	2.7 orders	HDR imaging,
	$L = \log_2(Y_{peak}) - \log_2(Y_{noise})$	9 f-stops	photography
signal to noise ratio	$SNR = 20 \cdot \log_{10}(Y_{peak}/Y_{noise})$	53[dB]	digital cameras

2.3   DYNAMIC RANGE

In principle, the term *dynamic range* is used in engineering to define the ratio between the largest and the smallest quantity under consideration. With respect to images, the observed quantity is the luminance level and there are several measures of dynamic range in use depending on the applications. They are summarized in Table 2.1.

The *contrast ratio* is a measure used in display systems and defines the ratio between the luminance of the brightest color it can produce (white) and the darkest (black). In case the luminance of black is zero, as for instance in HDR displays [2], the first controllable level above zero is considered as the darkest to avoid infinity. The ratio is usually normalized by the black level for clarity.

The *log exposure range* is a measure commonly adopted in high dynamic range imaging to measure the dynamic range of scenes. Here the considered ratio is between the brightest and the darkest parts of a scene given in luminance. The log exposure range is specified in orders of magnitude, which permits the expression of such ratios in a concise form using the logarithm base 10 and is usually truncated to one floating point position. It is also related to the measure of allowed exposure error in photography—exposure latitude. The *exposure latitude* is defined as the luminance range the film can capture minus the luminance range of the photographed scene and is expressed using logarithm base 2 with precision up to 1/3. The choice of logarithmic base is motivated by the scale of exposure settings, aperture closure (f-stops), and shutter speed (seconds), where one step doubles or halves the amount of captured light. Thus the exposure latitude tells the photographers how large a mistake they can make in setting the exposure parameters while still obtaining a satisfactory image. This measure is mentioned here, because its units, *f-stop steps* or *f-stops* in short, are often perhaps incorrectly used in HDR photography to define the luminance range of a photographed scene alone.

The *signal-to-noise ratio* (SNR) is most often used to express the dynamic range of a digital camera. In this context, it is usually measured as the ratio of the intensity that just saturates the image sensor to the minimum intensity that can be observed above the noise level of the sensor. It is expressed in decibel [dB] using 20 times base-10 logarithm.

The actual procedure to measure dynamic range is not well defined and therefore the numbers vary. For instance, display manufacturers often measure the white level and the black level with a separate set of display parameters that are fine-tuned to achieve the highest possible number which is obviously overestimated and no displayed image can show such a contrast. On the other hand, HDR images often have very few light or dark pixels. An image can be low-pass filtered before the actual dynamic range measure is taken to assure reliable estimation. Such filtering averages the minimum luminance thus gives a reliable noise floor, and smoothes single pixel with very high luminance thus gives a reasonable maximum amplitude estimate. Such a measurement is more stable compared to the non-blurred maximum and minimum luminance.

The last remaining aspect is the dynamic range that can be perceived by the human eye. The light scattering on the optic of the eye can effectively reduce the maximum luminance contrast that can be projected onto to retina to 2–3 log-10 units. However, since the eye is in fact a highly active sensor, which can rapidly change the gaze and locally adapt, people are believed to be able to perceive simultaneously the scenes of 4 or even more log-10 units [6, Section 6.2] of dynamic range.

CHAPTER 3

HDR Image and Video Acquisition

In recent years, several new techniques have been developed that are capable of capturing images with a dynamic range of up to 8 orders of magnitude at video frame rates. Such a range is practically sufficient to accommodate the full range of light present in the real world scenes. Together with the concept of the scene-referred representation of HDR contents this motivates that the HDR acquisition techniques output pixel intensities in well-calibrated photometric values. The varied techniques used in HDR capture require, however, careful characterization. In this chapter, we review the HDR capture techniques in the following section and describe the procedure for characterization of such cameras in terms of luminance in Section 3.2.

3.1 CAPTURE TECHNIQUES CAPABLE OF HDR

In principle, there are two major approaches to capturing high dynamic range: to develop new HDR sensors or to expose LDR sensors to light at more than one exposure level and later recombine these exposures into one high dynamic range image by means of a software algorithm. With respect to the second approach, the variation of exposure level can be achieved in three ways. The exposure can change in time, meaning that for each video frame a sequence of images of the same scene is captured, each with a different exposure. The exposure can change in space, such that the sensitivity to light of pixels in a sensor changes spatially and pixels in one image are non-uniformly exposed to light. Alternatively, an optical element can split light onto several sensors with each having a different exposure setting. Such software and hardware solutions to HDR capture are summarized in Sections 3.1.1–3.1.4.

3.1.1 Temporal Exposure Change

This is probably the most straightforward and the most popular method to capture HDR with a single low dynamic range sensor. Although such a sensor captures at once only a limited range of luminance in the scene, its operating range can encompass the full range of luminance through the change of exposure parameters. Therefore, a sequence of images, each exposed in such a way that a different range of luminance is captured, may together acquire the whole dynamic range of the scene, see Fig. 3.1. Such captures can be merged into one HDR frame by

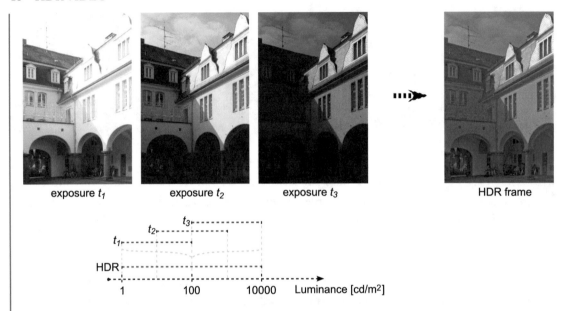

exposure t_1 exposure t_2 exposure t_3 HDR frame

FIGURE 3.1: Three consecutive exposures captured at immediate time steps t_1, t_2, t_3 contain different luminance ranges of a scene. The HDR frame merged from these exposures contains the full range of luminance in this scene. HDR frame tone mapped for illustration using a lightness perception inspired technique [14].

a simple averaging of pixel values across the exposures, after accounting for a camera response and normalizing by the exposure change [11, 12, 13] (for details on the algorithm refer to Section 3.2). Theoretically, this approach allows us to capture scenes of arbitrary dynamic range, with an adequate number of exposures per frame, and exploits the full resolution and capture quality of a camera.

HDR capture based on the temporal exposure change has, however, certain limitations especially in the context of video. Correct reconstruction of HDR from multiple images requires that each of the images captures exactly the same scene at a pixel level accuracy. This requirement cannot be practically fulfilled, because of camera motion and motion of objects in a scene, and pure merging techniques lead to motion artifacts and ghosting. To improve quality, such global and local displacements in images within an HDR frame must be realigned using for instance optical flow estimation. Furthermore, alignment of images that constitute one frame has to be temporarily coherent with adjacent frames. A complete solution that captures two images per frame and allows for real-time performance with 25 fps HDR video capture is described in [15]. An alternative solution that captures a much wider dynamic range of about 140 dB, but does not compensate for motion artifacts is available from [16].

e_3	e_0	e_3	e_0	e_3	e_0	e_3	e_0
e_2	e_1	e_2	e_1	e_2	e_1	e_2	e_1
e_3	e_0	e_3	e_0	e_3	e_0	e_3	e_0
e_2	e_1	e_2	e_1	e_2	e_1	e_2	e_1
e_3	e_0	e_3	e_0	e_3	e_0	e_3	e_0
e_2	e_1	e_2	e_1	e_2	e_1	e_2	e_1
e_3	e_0	e_3	e_0	e_3	e_0	e_3	e_0
e_2	e_1	e_2	e_1	e_2	e_1	e_2	e_1

scene capture without the mask   mask with optical densities varying per pixel   scene capture through the mask (varying pixel exposures)

FIGURE 3.2: Single exposure using a standard image sensor cannot capture full dynamic range of the scene (left). The mask with per pixel varying optical densities $e_3 = 4e_2 = 16e_1 = 64e_0$ (middle) can be put in front of a sensor. Using such a mask at least one pixel per four is well exposed during the capture (right). The right image is best viewed in the electronic version of the book.

The temporal exposure change requires a fast camera, because the effective dynamic range depends on the amount of captures per frame. For instance a 200 Hz camera is necessary to have a 25 fps video with 8 captures per frame that can give an approximate dynamic range of 140 dB [16]. With such a short time per image capture, the camera sensor must have a sufficiently high sensitivity to light to be able to operate in low light conditions. Unfortunately, such a boosted sensitivity usually increases noise.

3.1.2 Spatial Exposure Change

To avoid potential artifacts from motion in the scene, the exposure parameters may also change within a single capture [17], as an alternative to the temporal exposure change. The spatial exposure change is usually achieved using a mask which has a per pixel variable optical density. The number of different optical densities can be flexibly chosen and they can create a regular or irregular pattern. Nayar and Mitsunaga [17] propose to use a mask with a regular pattern of four different exposures as shown in Fig. 3.2. Such a mask can be then placed directly in front of a camera sensor or in the lens between primary and imaging elements.

For the pattern shown in Fig. 3.2, the full dynamic range can be recovered either by aggregation or by interpolation. The aggregation is performed over a small area which includes a capture of that area through each optical density, thus at several different exposures. The different exposures in the area are combined into one HDR pixel by means of a multi-exposure principle explained in the previous section, at the cost of reduced resolution of the resulting

HDR frame. To preserve the original resolution, HDR pixel values can also be interpolated from adjacent pixels in a similar manner as colors from the Bayer pattern. Depending on the luminance levels, aliasing and interpolation artifacts may appear.

The effective dynamic range in this approach depends on the number of different optical densities available in the pattern. A regular pattern of four densities, as shown in Fig. 3.2, such that $e_3 = 4e_2 = 16e_1 = 64e_0$ gives a dynamic range of about 85 dB for an 8-bit sensor [17]. The quantization step in the reconstructed HDR frame is non-uniform and increases for high luminance levels. The size of the step is, however, acceptable because it follows the gamma curve.

An alternative implementation of spatial exposure change, adaptive dynamic range imaging (ADRI), utilizes an adaptive optical density mask instead of a fixed pattern element [18,19]. Such a mask adjusts its optical density per pixel informed by a feedback mechanism from the image sensor. Thus saturated pixels increase the density of corresponding pixels in the mask, and noisy pixels decrease. The feedback, however, introduces a delay which can appear as temporal over- or under-exposure of moving high contrast edges. Such a delay, which is minimally one frame, may be longer if the mask with adapting optical densities has high latency.

Another variation of spatial exposure change is implemented in a sensor whose pixels are composed of more than one light sensing element each of which has a different sensitivity to light [20]. This approach is, however, limited by the size of the sensing element per pixel, and practically only two elements are used. Although in such a configuration, one achieves only a minor improvement in the dynamic range, so far only this implementation is applied in commercial cameras (Fuji Super CCD).

3.1.3   Multiple Sensors with Beam Splitters

Following the multi-exposure approach to extending dynamic range, one can capture several exposures per video frame at once using beam splitters [21,22]. The idea, so-called split aperture imaging, is to direct the light from the lens to more than one imaging sensor. Theoretically this allows us to capture HDR without making any quality trade-offs and without motion artifacts. In practice, however, the effective dynamic range depends on the number of sensors used in the camera and such a solution may become rather costly when a larger dynamic range is desired. Furthermore, splitting the light requires an increased sensitivity of the sensors.

3.1.4   Solid-State Sensors

There are currently two major approaches to extend the dynamic range of an imaging sensor. One type of sensor collects charge generated by the photo current. The amount of charge collected per unit of time is linearly related to the irradiance on the chip (similar to a standard

CCD chip [23]), the exposure time is however varying per pixel (sometimes called "locally auto-adaptive") [24,25,26]. This can for instance be achieved by sequentially capturing multiple exposures with different exposure time settings or by stopping after some time the exposure of the pixels that would be overexposed during the next time step. A second type of sensor uses the logarithmic response of a component to compute the logarithm of the irradiance in the analog domain. Both types require a suitable analog–digital conversion and typically generate a nonlinearly sampled signal encoded using 8–16 bits per pixel value. Several HDR video cameras based on these sensors are already commercially available. Such cameras allow us to capture dynamic scenes with high contrast, and compared to software approaches, offer considerably wider dynamic range and quality independent of changes in the scene content as frame-to-frame coherence is not required. The properties of two of such cameras: HDRC VGAx from IMS-CHIPS [27] and Lars III from Silicon Vision are studied in detail in Section 3.2.4.

3.2   PHOTOMETRIC CALIBRATION OF HDR CAMERAS

Ideally, in a photometrically calibrated system the pixel value output by a camera would directly inform about the amount of light that this camera was exposed to. However, in view of display-referred representation it has become important to obtain a visually pleasant image directly from a camera rather than such a photometric image. With the advance of high dynamic range imaging, however, the shift of emphasis in requirements can be observed. Many applications such as HDR video, capture of environment maps for realistic rendering, image-based measurements require photometrically calibrated images with absolute luminance values per pixel. For instance, the visually lossless HDR video compression (Chapter 5) is based on a model of human vision performance in observing differences in absolute luminance. An incorrect estimation of such performance due to the uncalibrated input may result in visible artifacts or less efficient compression. The capture technologies, however, especially in the context of HDR, are very versatile and a simple solution to obtain the photometric output from all types of cameras is not possible.

This section explains how to perform the absolute photometric calibration of HDR cameras and validates the accuracy of two HDR video cameras for applications requiring such calibration. For camera response estimation, an existing technique by Robertson et al. [28] is adapted to the specific requirements of HDR camera systems [29]. To obtain camera output in luminance units, the absolute photometric calibration is further determined. The achieved accuracy is estimated by comparing the measurements obtained with the absolute photometric calibration to measurements performed with a luminance meter and is discussed in the light of possible applications.

3.2.1   Camera Response to Light

An image or a frame of a video is recorded by capturing the irradiance at the camera sensor. At each pixel of the sensor, photons collected by a light sensitive area are transformed to an analog signal (electric charge) which is in turn read and quantized by a controller. Such a quantized signal is further processed to reduce noise, interpolate color information from the Bayer pattern, or enhance image quality, and is finally output from a camera. The camera response to irradiance, or light, describes the relation between incoming light and produced output value. The details of the capture process are often unknown thus the camera response is conveniently analyzed as a black box, which jointly describes the sensor response and built-in signal processing. In principle, the estimation of a camera response can be thought of as reading out the camera values for each single light quantity, although this is practically not feasible.

The camera response to light can be inversed to retrieve original irradiance value. Directly, the inverse model produces values that are only proportional (linearly related) to the true irradiance. The scale factor in this linear relation depends on the exposure settings and has to be estimated by additional measurements.

The HDR cameras have a nonlinear and sometimes non-continuous response to light and their output range exceeds 8 bit. Our choice of the framework for response estimation explained in the following section is motivated by its generality and the lack of restricting assumptions on the form of the response.

3.2.2   Mathematical Framework for Response Estimation

The camera response is estimated from a set of input images based on the expectation maximization approach [28]. The input images capture exactly the same scene, with correspondence at the pixel level, but the exposure parameters are different for each image. The exposure parameters have to be known and the camera response is observed as a change in the output pixel values with respect to a known change in irradiance. For the sake of clarity, in this section the exposure time is assumed to be the only parameter, but in general case it is necessary to know how many times more or less energy has been captured during each exposure. Since the exposure time is proportional to the amount of light captured in an image sensor, it serves well as the required factor. The mathematical formulas below follow the notation given in Table 3.1 and consider only images with one channel.

There are two unknowns in the estimation process. The primary unknown, the camera response function I, models the relation between the camera output values and the irradiance at the camera sensor, or luminance in the scene. The camera output values for a scene are provided as input images, but the irradiance x coming from the scene is the second unknown. The estimation process starts with an initial guess on the camera response function, which for instance can be a linear response, and consists of two steps that are iterated. First, the irradiance

TABLE 3.1: Symbols and notation in formulas for response estimation

i—image index

j—pixel position index

t_i—exposure time of image i

y_{ij}—pixel value of input image i at position j

$I(\cdot)$—camera response function

x_j—estimated irradiance at pixel position j

$w(\cdot)$—weighting function from certainty model

m—pixel value from a set of possible camera output values

from the scene is computed from the input images based on the currently estimated camera response. Second, the camera response is refined to minimize the error in mapping pixel values from all input images to the computed irradiance. The process is terminated when the iteration step no longer improves the camera response. The details of the process are explained below.

Estimation of Irradiance

Assuming that the camera response function I is correct, the pixel values in the input images are mapped to the relative irradiance by using the inverse function I^{-1}. Such relative irradiance is proportional to the true irradiance from the scene by a factor influenced by the exposure parameters (e.g., exposure time), and the mapping is called linearization of camera output. The relative irradiance is further normalized by the exposure time t_i to estimate the amount of energy captured per unit of time in the input images i at pixel position j:

$$x_{ij} = \frac{I^{-1}(y_{ij})}{t_i}. \tag{3.1}$$

Each of the x_i images contains a part of the full range of irradiance values coming from the scene. This range is determined by the exposure settings and is limited by the dynamic range of the camera sensor. The complete irradiance at the sensor is estimated from the weighted average of this partial captures:

$$x_j = \frac{\sum_i w_{ij} \cdot x_{ij}}{\sum_i w_{ij}}. \tag{3.2}$$

The weights w_{ij} are determined for camera output values by the certainty model discussed later in this section. Importantly, the weights for the maximum and minimum camera output values are equal to 0, because the captured irradiance is bound to be incorrect in the pixels for which the sensor has been saturated or captured no energy.

Refinement of Camera Response

Assuming that the irradiance at the sensor x_j is correct, one can recapture the camera output values y'_{ij} in each of the input images i by using the camera response:

$$y'_{ij} = I(t_i \cdot x_j). \tag{3.3}$$

In the ideal case when the camera response I is perfectly estimated, the y'_{ij} is equal to y_{ij}. During the estimation process, however, the camera response function needs to be optimized for each camera output value m by averaging the recaptured irradiance x_j for all pixels in the input images y_{ij} that are equal to m:

$$E_m = \{(i, j) : y_{ij} = m\}, \tag{3.4}$$

$$I^{-1}(m) = \frac{1}{\text{Card}(E_m)} \sum_{i,j \in E_m} t_i \cdot x_j. \tag{3.5}$$

The Certainty Model

The presence of noise in the capture process is conveniently neglected in the capture model in Eqs. (3.1), (3.3). A complete capture model would require characterization of possible sources of noise and incorporation of appropriate noise terms to the equation. This would require further measurements and analysis of particular capture technology in the camera, thus is not practical. Instead, the noise term can be accounted for by an intuitive measure of confidence in the accuracy of captured irradiance. In typical 8-bit cameras, for instance, one would expect high noise in the low camera output values, quantization errors in the high values, and good accuracy in the middle range. An appropriate certainty model can be defined by the following Gaussian function:

$$w(m) = \exp\left(-4 \cdot \frac{(m - 127.5)^2}{127.5^2}\right). \tag{3.6}$$

The certainty model can be further extended with knowledge about the capture process. Normally, longer exposure times, which allow us to capture more energy, tend to exhibit less random noise than short ones. Therefore, an improved certainty model for input images y_{ij} can be formulated as follows:

$$w_{ij} = w(y_{ij}) \cdot t_i^2. \tag{3.7}$$

Such weighting function minimizes the influence of noise on the estimation of irradiance in Eq. (3.2). This happens apart from noise reducing properties of the image averaging process itself.

Minimization of Objective Function

After the initial assumption on the camera response I, which is usually linear, the response is refined by iteratively computing Eqs. (3.2) and (3.5). At the end of every iteration, the quality of estimated camera response is measured with the following objective function:

$$O = \sum_{i,j} w(y_{ij}) \cdot (I^{-1}(y_{ij}) - t_i \cdot x_j)^2. \qquad (3.8)$$

The objective function measures the error in the estimated irradiance for input images y_{ij} when compared to the simulated capture of the true irradiance x_j. The certainty model requires that the camera output values in the range of high confidence give more accurate irradiance estimates. The estimation process is terminated as soon as the objective function O falls below the predetermined threshold.

The estimation process requires an additional constraint, because two dependent unknowns are calculated simultaneously. Precisely, the values of x_j depend on the mapping of I and the equations are satisfied by infinitely many solutions to I which differ by a scale factor. Convergence to one solution is enforced, in each iteration, through normalization of the inverse camera response I^{-1} by the irradiance causing the medium camera output value $I^{-1}(m_{\mathrm{med}})$.

3.2.3   Procedure for Photometric Calibration

In the following sections, a step-by-step procedure for photometric calibration of HDR cameras is outlined.

Scene Setup for Calibration

The response estimation algorithm requires that each camera output value is observed in more than one input image. Moreover, frequent observations of the value reduce the impact of noise. Therefore, an ideal scene for calibration is static, contains a range of luminance wider than the expected dynamic range of the camera, and smoothly changing illumination which gives a uniform histogram of output values. Additionally, neutral colors in the scene can minimize the possible impact of color processing in a color camera.

When calibrating HDR cameras, a static scene with a sufficiently wide dynamic range may not be feasible to create. In such a case, it is advisable to prepare several scenes, each covering a separate but partially overlapping luminance range, and stitch them together into a single image.

Capture of Images for Calibration

Input images for the calibration process capture exactly the same scene with varying exposure parameters. A steady tripod and remote control of a camera are essential requirements. A slight out-of-focus reduces edge aliasing due to sensor resolution and limits potential sharpening in a camera, thus makes the estimation process more stable.

HDR cameras often do not offer any adjustment of exposure parameters or available adjustments are not bound to have a linear influence on captured energy. The aperture value cannot be changed to adjust the exposure because it modifies the depth-of-field, vignetting, and diffraction pattern, thus practically changes the scene between input images. Instead, the optical filters, such as neutral density (ND) filters, can be mounted in front of the lens to limit the amount of irradiance at the sensor at a constant exposure time. The ND filters are characterized by their optical density which defines the amount of light attenuation in logarithmic scale. In the response estimation framework, such optical density can be used to calculate a simulated exposure time of captured images:

$$t_i = t_0 \cdot 10^{D_i}, \tag{3.9}$$

where t_i is simulated exposure time of image i captured through an optical filter of density D_i calculated with respect to the true exposure time t_0. If t_0 is not known from the camera specifications, it can be assumed equal to 1. One should make sure that the optical filters are spatially uniform and equally reduce the intensity of all captured wavelengths.

Following the analysis in [30], it can be suggested to acquire two images that are exposed similarly and one that is considerably different. Additionally, when calibrating a video camera one may capture a larger number of frames for each of the exposures. Such a superfluous number of input images will reduce the influence of image noise on the response estimation.

Absolute Photometric Calibration

The images of the calibration scene are input to the estimation framework from Section 3.2.2 to obtain a camera response. For an RGB or multi-spectral camera, the camera response has to be estimated for each color channel separately. Here, a camera that captures monochromatic images with spectral efficiency corresponding to luminance is assumed. In the case of an RGB camera, an approximation of luminance Y can be calculated from color channels using RGB to XYZ color transform.

The relative luminance values obtained from the estimated response curve are linearly proportional to the absolute luminance with a scale factor dependent on the exposure parameters and the lens system. Absolute calibration is based on the acquisition of a scene containing patches with known luminance Y. The scale factor f is determined by minimizing relative error between known and captured luminance values:

$$Y = f \cdot I^{-1}(m). \tag{3.10}$$

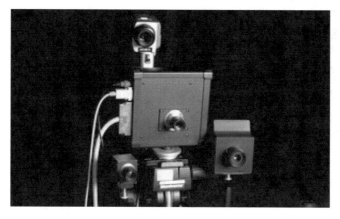

FIGURE 3.3: Cameras used in our experiment: HDRC VGAx (lower left), Silicon Vision Lars III (center), Jenoptik C14 (lower right), and Minolta LS-100 luminance meter (top).

3.2.4   Example Calibration of HDR Video Cameras

The photometric calibration is demonstrated in this section on two HDR video cameras: the Silicon Vision Lars III camera and the HDRC VGAx camera. The Jenoptik C14, a high-end, CCD based LDR camera (see Fig. 3.3), is also included for comparison purposes. The Lars III sensor is an example of a locally auto-adaptive image sensor [26]: the exposure is terminated for each individual pixel after one out of 12 possible exposure times (usually powers of 2). For every pixel, the camera returns the amount of charge collected until the exposure was terminated as a 12-bit value and a 4-bit time-stamp. The HDRC sensor is a logarithmic-type sensor [31] and the camera outputs 10-bit values per pixel [27].

Estimation of Camera Response

To cover the expected dynamic range of calibrated cameras, in the presented case it was necessary to acquire three scene setups with varied luminance characteristic (see Fig. 3.4): a scene with moderate illumination, the same scene with a strong light source, and a light source with reflector shining directly toward the cameras. Stitching these three images together yields an input for the response estimation algorithm covering a dynamic range of more than 8 orders of magnitude. Each scene setup has been captured without any filter and with a $\times 1.5$ ND filter and a $\times 10$ ND filter. The response of C14 camera was estimated using a series of 13 differently exposed images of a GretagMacbeth ColorChecker.

The estimated responses of the three cameras are shown in Fig. 3.5. The certainty functions have been modeled using Eq. (3.6) such that maximum confidence is assigned to the middle of operational luminance range and limits to zero at the camera output levels dominated by noise. A single response curve has been estimated for the monochromatic Lars III camera

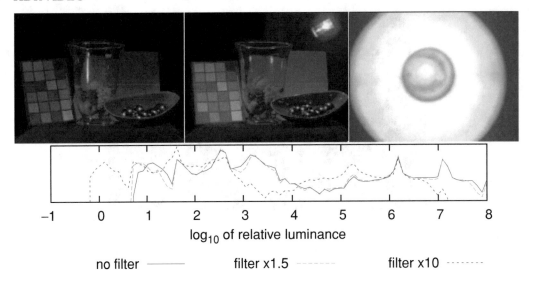

FIGURE 3.4: Three scene setups for the estimation of response curves (tone mapped for presentation). The histogram shows the luminance distribution in the stitched images for acquisition without filter, and using ND filters with ×1.5 and ×10 optical density. This setup covers 8 orders of luminance magnitude.

and separate curves have been determined for the three color channels of the other cameras. As the raw sensor values of the HDRC camera before Bayer interpolation have been available, the response curve for each channel has been directly estimated from corresponding pixels in order to avoid possible interpolation artifacts.

Figure 3.5 shows that the response curves of the two HDR cameras both cover a considerably wider range of luminance than the high-end LDR camera that covers a range of about 3.5 orders of magnitude. The different shapes of the HDR response curves are caused by their respective sensor technology and the encoding. The logarithmic HDRC VGAx camera has the highest dynamic range (more than 8 orders of magnitude), but an offset in the A/D conversion makes the lower third of the 10-bit range unusable. The multiple exposure values of the locally auto-adaptive Lars III camera are well visible as discontinuities in the response curve. Note that the luminance range is covered continuously and gaps are only caused by the encoding. The camera covers a dynamic range of about 5 orders of magnitude. Noise at the switching points between exposure times is well visible.

Results of Photometric Calibration

The inverse of the estimated responses converts the camera output values into relative luminance values. To perform an absolute calibration, the GretagMacbeth ColorChecker chart has been acquired under six different illumination conditions. The luminance of the gray patches was

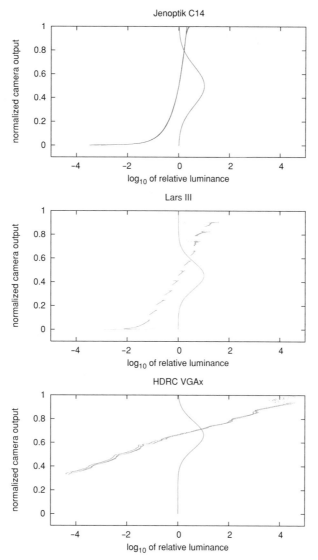

FIGURE 3.5: The estimated response curves and corresponding weighting functions from the certainty model (value 1.0 represents the full confidence in capture accuracy, 0.0 represents no confidence). The peaks of the weighting functions are centered at the middle of the operational range of each camera.

measured using a Minolta LS-100 luminance meter yielding a total of 36 samples and an optimal scale factor was determined for each camera. The accuracy of the absolute calibration for the 36 patches can be seen in Fig. 3.6. The calibrated camera luminance values are well aligned to the measured values proving that the response curve recovery was accurate. The

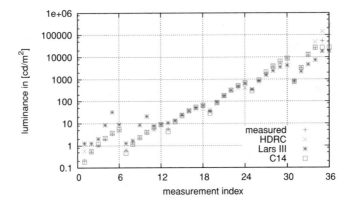

FIGURE 3.6: The results of absolute calibration. The estimated response curves were fitted to the measurements of six gray patches of GretagMacbeth ColorChecker chart under six different illumination conditions.

average relative error for these data points quantifies the quality of the absolute calibration. For the HDRC camera, relative error in the luminance range of $1-10\,000$ cd/m^2 is 13% while the relative error for the Lars III camera in the luminance range of $10-1000$ cd/m^2 amounts to 9.5%. Note that these results can be obtained with a single acquisition. Using multiple exposures, the C14 camera is capable of an average relative error of below 7% in the range $0.1-25\,000$ cd/m^2, thus giving the most accurate results.

3.2.5   Quality of Luminance Measurement

The described procedure for photometric calibration of HDR cameras proved to be successful; however, the accuracy obtained for example HDR cameras is not very high. Although one should not expect to match the measurement quality of a luminance meter, still the relative error of the LDR camera is lower than of HDR cameras. Besides, both HDR cameras keep the error below 10% only in the range of luminance that is much narrower than their operational range. The low accuracy in low illumination is mostly caused by noise in the camera and can be hardly improved in the calibration process. On the other hand, the low accuracy in high luminance range can be affected by the calibration process: a very bright scene was required to observe high camera output values. The only possibility of getting a bright enough scene was to directly capture a light source, but the intensity of the light source might not have been stable during the capture and an additional noise has been introduced to the estimation process.

To improve the results, the estimated response can be fit to an *a priori* function appropriate for the given HDR sensor. Thus, for the HDRC camera the parameters of a logarithmic function $y_j = a * \log(x_j) + b$ are fit and for the decoded values[1] of the Lars III camera a linear function $y_j = a * x_j + b$ is used. The relative errors achieved by the pure response estimation including absolute calibration and the function fit are compared in Fig. 3.7. The average relative error is equal to about 6% for the HDRC camera and luminance values above $1 \, \text{cd/m}^2$. For the Lars III camera, it is also about 6% for luminance values above $10 \, \text{cd/m}^2$. Especially for high luminance values above $10\,000 \, \text{cd/m}^2$, the calibration via function fitting provides more accurate results. In addition, the fitting approach allows us to extrapolate the camera response for values beyond the range of the calibration scene. To verify this, an extremely bright patch ($194\,600 \, \text{cd/m}^2$ in the presented case) can be acquired using the calibrated response of the HDR cameras and compared to the measurement of the light meter. Only the readout from the HDRC camera derived via function fitting is reliable while the HDRC response curve seems to be bogus in that luminance range. The Lars III camera reached the saturation level and yielded arbitrary results. Likewise, this patch could not be recorded with the available settings of the LDR camera.

3.2.6   Alternative Response Estimation Methods

In principle, three different approaches can be used to estimate the response of 8-bit cameras ([6] provides a good survey, [32] gives a theoretical account of ambiguities arising in the recovery of camera response from images taken at different exposures). The method of Robertson et al. [28] has been selected because of its unconstrained applicability to varied types of sensors in cameras. For completeness, the remaining two methods are briefly discussed in view of possible application to photometric calibration of HDR cameras.

The algorithm developed by Debevec and Malik [12] is based on the concept that a particular pixel exposure is defined as a product of the irradiance at the film and the exposure time, transferred by the camera response function. This concept is embedded in an objective function which is minimized to determine the camera response curve. The objective function is additionally constrained by the assumption that the response curve is smooth, which is essential for the minimization process. Whereas this assumption is generally true for LDR cameras based on CCD technology, the response curve is normally not smooth in locally autoadaptive HDR sensors. Furthermore, the process of recovering the response curve is based on solving a set of linear equations. While the size of the matrix representing these linear equations is

---

[1] According to the data sheet, the 16-bit output value of Lars III camera is in fact a composite of a 12-bit mantissa m and a 4-bit exponent value e; i.e. $y_j = m \cdot 2^e$.

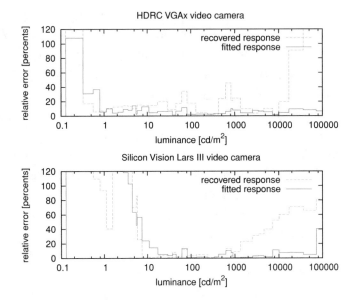

FIGURE 3.7: Comparison of the relative errors in luminance measurement achieved by the pure response estimation including absolute calibration and by the function fit.

reasonable for 8-bit data, memory problems may occur for arbitrary precision data typical to HDR acquisition so that extensive sub-sampling is required.

The method proposed by Mitsunaga and Nayar [13] computes a radiometric response function approximated using a high-order polynomial without precise knowledge of the exposures used. The refinement of the exposure times during the estimation process is major advantage; however, the process itself is limited to computation of the order of the polynomial and its coefficients. The authors state that it is possible to represent virtually any response curve using a polynomial. This fact is true for LDR cameras based on a CCD sensor; however, it is not possible to approximate the logarithmic response of some CMOS sensors in this manner. Polynomial approximation also assumes that the response curve is continuous, which depends on the encoding.

Grossberg and Nayar [32] show how the radiometric response function can be related to the histograms of non-registered images with different exposures. This enables to deal with the scene and camera motion while the images are captured, under the condition that the distribution of scene radiance does not change significantly between images.

3.2.7   Discussion

The ability to capture HDR data has a strong impact on various applications because the acquisition of dynamic sequences that can contain both very bright and dark luminance (such as sun and deep shadows) at the same moment is unprecedented. Photometrically calibrated HDR contents offer further benefits. Perceptually enabled algorithms employed in compression or tone mapping can appropriately simulate the behavior of human visual system. Dynamic environment maps can be captured in real time to faithfully convey the illumination conditions of the real world to rendering algorithms. Some of such applications are discussed in Section 9.2. The results of global illumination solutions can than be directly compared to the real-world measurements as illustrated in Fig. 9.1 in Section 9.1. The calibrated HDR video cameras can further increase the efficiency of measuring appearance of complex materials in the form of bi-directional reflectance distribution function (BRDF), Section 9.2.2.

With respect to the presented calibration methods, while the relative error achieved by the function fitting approach is lower, the response estimation algorithm is useful to obtain the exact shape of the camera response and to give confidence that the chosen *a priori* function is correct. It can also help to understand the behavior of the sensor, especially if the encoding is unknown. The low precision of the measurements in the luminance range below $10 \, \text{cd/m}^2$ is a clear limitation which can be explained by the high noise level in the sensors. The quality of a high-end CCD camera such as the Jenoptik C14 combined with traditional HDR recovery algorithms still cannot be achieved consistently over the whole dynamic range of the HDR cameras.

The function fitting approach has strong advantages in the quality of the results and the ability to extrapolate from the calibration data. The confidence in extrapolated measurements is however limited and the error cannot be predicted because the exact shape of the response function in this range is unknown. Finally, the accuracy of the photometric calibration is not the only important quality measure. Depending on the application, other issues such as the quantization of the luminance values might have an important influence on the quality of the measurements and need to be further investigated.

In Chapter 10, we provide more information on the *pfscalibration* software package [33], which can be used for photometric calibration of both LDR and HDR cameras. The package is available under the URL:

```
http://www.mpi-inf.mpg.de/resources/hdr/calibration/pfs.html
```

CHAPTER 4

HDR Image Quality

The performance of many imaging algorithms, such as image compression, is often a function of visual quality. The visual quality can be most reliably measured in subjective studies, in which a group of people assigns quality scores to the presented video or images. Such studies, however, are both tedious and expensive and often result in high variance between observers. In many areas, it is much more practical to use instead objective quality metrics, which can estimate perceived quality without subjective judgements. This chapter gives a short classification of the available metrics and describes in more detail a metric designed for comparing high dynamic range images.

4.1 VISUAL METRIC CLASSIFICATION

Although numerous image comparison algorithms are classified as quality metrics, it does not mean that they compute the same quality measure. Some metrics are better suited for estimating quality of low-bandwidth video transmission, where large distortions are common and acceptable, and other for compression of medical images, where visual distortions must be avoided. Therefore, it is important to distinguish between all kinds of visual metrics and choose the one that is appropriate for a particular application.

A high-level classification of the visual metrics is shown in Fig. 4.1. Depending whether a metric requires a non-distorted reference image, some limited statistics of such an image or no image at all, it can be classified as a full-reference, limited-reference, and no-reference. Although there are extensive studies on the limited-reference and no-reference metrics, majority of quality metrics require a reference image. No-reference metrics are usually limited to a single type of distortion, such as JPEG blocky artifacts or blurring, and cannot match in accuracy the full-reference metrics.

The simplest kind of the full-reference metrics are arithmetical measures, such as the peak-signal-to-noise ratio (PSNR) or mean-squared error (MSE). Despite their simplicity and known cases when they fail, these are the most commonly used metrics in estimating performance of video compression. In fact the PSNR can give quite accurate estimates of quality for video compression, comparable with much more complex perceptually weighted metrics,

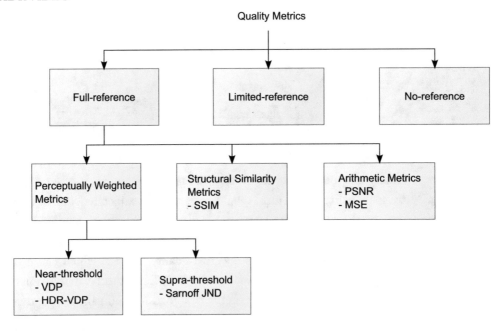

FIGURE 4.1: Classification of quality metrics.

mostly because video compression itself is driven by visual models. The structural similarity metrics, such as SSIM [34], offer a trade-off between a complexity of the perceptually-weighted metrics and the simplicity of the arithmetic metrics. They combine local statistical measures of an image to compute a quality estimate that achieves a good correlation with the quality measures found in subjective studies.

The potentially most accurate metrics are those that model the human visual system to predict perceivable distortions. Most of them are quite accurate at predicting just noticeable distortions which are near the discrimination threshold of the human visual system. The near-threshold metrics, such as VDP [35] or HDR-VDP, can quite precisely predict whether a human observer will spot any difference between two images shown, but they cannot make a difference between the distortions that are far above the threshold. For example, they make little distinction between poor and extremely poor quality video. This task is more suitable for the supra-threshold metrics, which can estimate not only presence, but also the magnitude of distortion [36, 37].

The metrics can be further divided into those that produce a single quality measure (e.g., a numerical value) for an image or a video sequence and those that produce a distortion map, which estimates the local magnitude of distortion or probability of detection (usually for each pixel). The performance of a metric that computes a single quality measure is usually evaluated in comparison with the subjective data, for example from the LIVE image quality assessment database [38].

This chapter does not cover the area of quality metrics in general, but focus on a particular metric designed especially for high dynamic range images.

4.2 A VISUAL DIFFERENCE PREDICTOR FOR HDR IMAGES

Most of the objective quality metrics have been designed to operate on video and images that are to be displayed on CRT or LCD displays. While this assumption seems to be clearly justified in the case of low dynamic range images, it poses problems as new applications that operate on HDR data become more common. A perceptual HDR quality metric could be used for the validation of the HDR image and video encodings. Another application may involve steering the computation in a realistic image synthesis algorithm, where the amount of computation devoted to a particular region of the scene would depend on the visibility of potential artifacts.

The HDR-VDP extends a well-known visual difference predictor [35] to better cope with high contrast images and a broad range of luminance conditions. The extensions focus on the accurate modeling of the visibility threshold under the assumption that an observer can locally adapt to luminance levels of a scene. This makes the predictor more conservative but also more reliable when scenes with significant differences of luminance are analyzed. Such local adaptation is essential for a good reduction of contrast visibility in HDR images, as a single HDR image can contain both dimly illuminated interior and strong sunlight.

The data-flow diagram of the HDR-VDP is shown in Fig. 4.2. The HDR-VDP receives a pair of images as an input (original and distorted, for example by image compression) and generates a map of probability values, which indicates how likely the differences between those two images are perceived. Both images should be scaled in the units of luminance. In the case of low-dynamic range images, pixel values should be inverse gamma corrected and calibrated according to the maximum luminance of the display device. In the case of HDR images no such processing is necessary, however luminance should be given in cd/m^2.

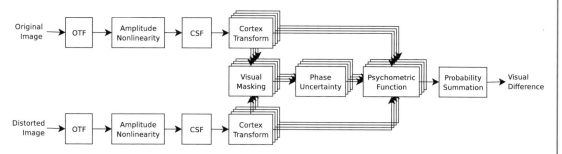

FIGURE 4.2: Data-flow diagram of the high dynamic range visible difference predictor (HDR-VDP).

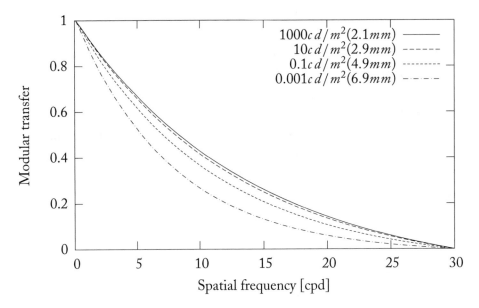

FIGURE 4.3: Optical MTFs from the model of Deeley et al. [39] for different levels of adaptation to luminance and pupil diameters (given in parentheses).

The first three stages of HDR-VDP model behavior of the optics and retina. Both a reference and a test images are filtered by the optical transfer function (OTF), which simulates light scattering in the cornea, lens, and retina. The OTF used in the HDR-VDP is shown in Fig. 4.3. Figure 4.4 demonstrates the effect of the OTF on an HDR image with a relatively bright regions. HDR images can contain high luminance objects (sun, lamps, brightly illuminated windows) that can significantly affect contrast perception in the neighboring regions.

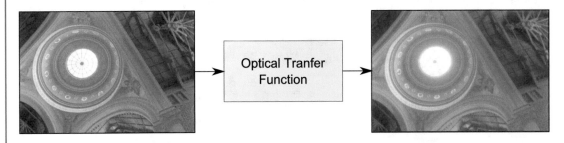

FIGURE 4.4: The result of filtering and image with the optical transfer function (OTF) of the human eye. The *Memorial Church* image courtesy of Paul Debevec.

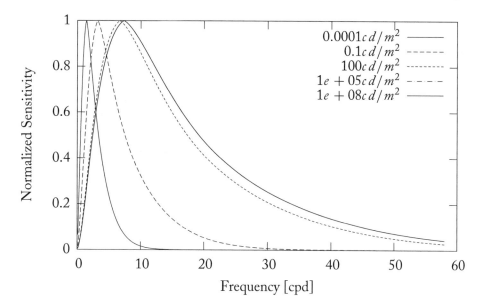

FIGURE 4.5: Family of normalized Contrast Sensitivity Functions (CSF) for different adaptation levels. The peak sensitivity shifts toward lower frequencies as the luminance of adaptation decreases.

To account for the nonlinear response of photoreceptors to light, the amplitude of the signal is nonlinearly compressed and expressed in the units of just noticeable differences (JND). Such nonlinearity is very similar to the JND-encoding discussed in Section 5.1.6, but is derived from the contrast sensitivity function (CSF), used in the next processing step. Because the HVS is less sensitive to low and high spatial frequencies, in the next step a JND-scaled image is filtered by the CSF. Unlike the original VDP, the HDR-VDP locally adapts the CSF filtering kernel depending on the adaptation luminance. The adaptation luminance shifts the CSF both horizontally and vertically. Since the vertical shifts affecting the peak contrast sensitivity are already modeled by the amplitude nonlinearity, the CSF is normalized so that the peak has value 1, and only horizontal shifts must be taken into account. The horizontal shifts of the CSF due to adaptation luminance are shown in Fig. 4.5.

The OTF, amplitude nonlinearity, and the CSF filtering steps are mostly responsible for contrast reduction in the HVS. The next two computational blocks, the cortex transform and visual masking, decompose the images into spatial and orientational channels and predict perceivable differences in each channel separately. Phase uncertainty further refines the prediction of masking by removing dependence of masking on the phase of the signal. In the final error pooling stage, the probabilities of visible differences are summed up for all channels and a map of detection probabilities is generated.

4.2.1 Implementation

The source code of HDR-VDP is available under the GPL license and can be downloaded from the web page `http://hdrvdp.sourceforge.net/`. It is integrated with *pfstools* package (refer to Chapter 10), which can read most of the HDR file formats. The software provides a ready-to-use metric that can be used in a broad range of digital imaging applications, ranging from validation of computer graphics algorithms to detection of artifacts in compressed images.

The detailed documentation of the HDR-VDP software can be found on the web page. To give an impression how the software operates, the box below shows a typical usage scenario:

```
vdp␣original.exr␣distorted.exr␣prediction.png
```

Predict differences between an original `original.exr` and distorted `distorted.exr` images and create the visualization of the prediction in `prediction.png`.

CHAPTER 5

HDR Image, Video, and Texture Compression

The bit-depth precision of majority of image and video formats can soon become insufficient for the new generation of displays. The traditional image and video formats, such as JPEG, PNG, or MPEG, employ color spaces that fail to represent scenes of dynamic range over 2 or 3 orders of magnitude and extended color gamut. The 8-bit-per-color-channel encoding was more than sufficient when such formats were designed, and the best CRT displays could achieve contrast ratio of 1:200 and their peak luminance did not exceed 100 cd/m^2. Now, commercially available displays can show contrast of 1:3000[1]. The prototypes of HDR displays are capable of showing contrast 1:200 000 and have the peak luminance of 3000 cd/m^2(refer to Section 7.2). Moreover, the improvements of LED display backlight make it possible to achieve more saturated colors and thus wider color gamut. These new advances in display technology make essential that video and image compression formats are extended to support new displays.

Despite the diversity of display technologies (LCD, Plasma, DLP, etc.), the most popular image and video file formats are still device dependent. The gamma correction nonlinearity, employed in most color spaces used for compression, was originally designed for the CRT displays [40]. When technology changes rapidly, developing standards based on the characteristics of the particular type of devices does not seem to be appropriate.

In typical imaging pipelines, it is commonly assumed that the decoded images or video are directly displayed. As the complexity and diversity of displays increase, it can be expected that the future displays will employ additional rendering step, in which the dynamic range and color gamut is reduced to match the display capabilities (tone mapping), the content is adapted to the viewing conditions (different rendering for bright and dark room), additional effects and enhancements are applied. Figure 5.1 demonstrates some effects that simulate the human visual system or a camera, that can be added in real-time to the video stream [41].

High dynamic range (HDR) imaging is a very attractive way of capturing real-world appearance, since it assumes the preservation of complete and accurate luminance (or spectral

---

[1]For a single frame, as of 2007.

FIGURE 5.1: A range of perceptual effects that can be simulated based on HDR data. From left to right: visual glare (see light scattering at the edges of the objects); motion blur can be correctly simulated in linear luminance domain (right half); given absolute luminance values, color deficiency of night (scotopic) vision can be simulated. The source images courtesy of Paul Debevec, Spheron VR, and vr architects.

radiance) values that can be found in a scene. Each pixel is represented as a triple of floating point values, which can range from 10^{-5} to 10^{10}. Such a huge range of values is dictated by both real-world luminance levels and the capabilities of the human visual system (HVS), which can adapt to a broad range of luminance levels, ranging from scotopic ($10^{-5} - 10$ cd/m^2) to photopic ($10 - 10^6$ cd/m^2) conditions. Obviously, floating point representation results in huge memory and storage requirements and is impractical for storage and transmission of images and video. Therefore, better techniques of encoding HDR pixel values are discussed in Section 5.1.

This chapter is intended to give an overview of the current state-of-the-art in the high-fidelity image, video, and texture coding. Section 5.2 gives an overview of the image and Section 5.3 of the video formats that are intended to preserve higher fidelity. As HDR formats have just started gaining popularity, it is important to provide backward-compatibility with the existing LDR formats. The schemes for backward-compatible compression of HDR images and video are described in Section 5.4. Finally, Section 5.5 reviews some recent texture compression schemes.

5.1   HDR PIXEL FORMATS AND COLOR SPACES

Choice of the color space and the pixel encoding used for image or video compression has a great impact on the compression performance and capabilities of the encoding format. While representing pixel values as a triple of 32-bit floating point numbers gives more than sufficient precision and good flexibility in data processing, such encoding does not use memory efficiently and is not compatible with most image and video compression standards. For this reason,

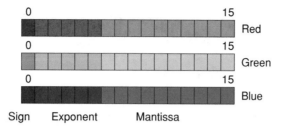

FIGURE 5.2: Red–green–blue component encoding using half-precision floating point numbers.

several HDR pixel encoding and color spaces are used in popular HDR image formats. This section gives an overview of these pixel encodings.

5.1.1   Minifloat: 16-Bit Floating Point Numbers

Graphics cards from nVidia and ATI can use more compact representation for floating point numbers, known as *half-precision float, fp16*, or *S5E10*. The S5E10 indicates that the floating point number consist of one bit of sign, 5-bit exponent, and 10-bit mantissa, as shown in Fig. 5.2. Such 16-bit floating point formats is also used in the OpenEXR image format (see Section 5.2.2).

The half-precision float offers flexibility of the floating point numbers at the half storage cost of the typical 32-bit floating point format. Floating point numbers are well suited for encoding linear luminance and radiance values, as they can easily encompass large dynamic ranges. One caveat of the half-precision float format is that it can represent numbers up to the maximum value 65 504, which is less than for instance luminance of bright light sources. For this reason, the HDR images given in absolute luminance or radiance units often need to be scaled down by a constant factor before storing them in the half-precision float format.

5.1.2   RGBE: Common Exponent

The *RGBE* pixel encoding is used in the Radiance file format, which will be discussed in Section 5.2.1. The RGBE pixel encoding represents colors using four bytes: the first three bytes encode red, green, and blue color channels, and the last byte is a common exponent for all channels (see Fig. 5.3). RGBE is essentially a custom floating point representation of pixel values, which uses 8 bits to represent exponent and another 8 bits to represent mantissa (8E8). RGBE encoding takes advantage of the fact that all color channels are strongly correlated in the RGB color spaces and their values are at least of the same order of magnitude. Therefore, there is no need to store a separate exponent for each color channel.

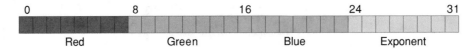

FIGURE 5.3: 32-bit per pixel RGBE encoding.

The conversion between from (R, G, B, E) bytes to red, green, and blue trichromatic color values (r, g, b) is done using the formulas:

$$(r, g, b) = \begin{cases} \dfrac{(R, G, B) + 0.5}{256} \, 2^{E-128} \, \dfrac{\text{exposure}}{E_w} & \text{if } E \neq 0 \\ (0, 0, 0) & \text{if } E = 0 \end{cases} \qquad (5.1)$$

where per image "exposure" parameter can be used to adjust absolute values, and E_w is the efficacy of the white constant equal to 179. Both these terms are used in the Radiance file format but are often omitted in other implementations.

The inverse transformation is given by

$$E = \begin{cases} \lceil \log_2 (\max\{r, g, b\}) + 128 \rceil & \text{if } (r, g, b) \neq 0 \\ 0 & \text{if } (r, g, b) = 0 \end{cases} \qquad (5.2)$$

$$(R, G, B) = \left\lfloor \frac{256 \, r}{2^{E-128}} \right\rfloor$$

where $\lceil \cdot \rceil$ denotes rounding up to the nearest integer and $\lfloor \cdot \rfloor$ rounding down to the nearest integer.

5.1.3   LogLuv: Logarithmic Encoding

One shortcoming of floating point numbers is that they are not optimal for image compression methods. This is partly because additional bits are required to encode mantissa and exponent separately, instead of a single integer value. Such representation, although flexible, is not necessary for color data. Furthermore, precision error of floating point numbers varies across the full range of possible values and is different from the "precision" of our visual system. Therefore, better compression can be achieved when integer numbers are used to encode HDR pixels.

The *LogLuv* pixel encoding [42] requires only integer numbers to encode the full range of luminance and color gamut that is visible to the human eye. It is an optional encoding in the TIFF library. This encoding benefits from the fact that the human eye is not equally sensitive to all luminance ranges. In the dark, we can see a luminance difference of a fraction of 1 cd/m^2, while in the sunlight we need a difference of tens of cd/m^2 to see a difference. This effect is often called luminance masking. But if, instead of luminance, a logarithm of luminance is considered,

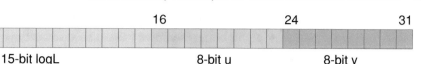

Sign   15-bit logL   8-bit u   8-bit v

FIGURE 5.4: 32-bit per pixel LogLuv encoding.

the detectable threshold values do not vary so much and a constant value can be a plausible approximation of the visible threshold. Therefore, if a logarithm of luminance is encoded using integer numbers, quantization errors roughly correspond to the visibility thresholds of the human visual system, which is a desirable property for pixel encoding.

The 32-bit LogLuv encoding uses two bytes to encode luminance and another two bytes to represent chrominance (see Fig. 5.4). Chrominance is encoded using the CIE 1976 Uniform Chromacity Scales u' v':

$$u' = \frac{4X}{X + 15Y + 3Z} \qquad v' = \frac{9Y}{X + 15Y + 3Z} \qquad (5.3)$$

which can be encoded using 8-bits:

$$u_{8bit} = u' \cdot 410 \qquad v_{8bit} = v' \cdot 410. \qquad (5.4)$$

Note that the u' and v' chromatices are used rather than u^* and v^* of the $L^*u^*v^*$ color space. Although u^* and v^* give better perceptual uniformity and predict loss of color sensitivity at low light, they are strongly correlated with luminance. Such correlation is undesired in image or video compression. Besides, the u^* and v^* chromatices could reach high values for high luminance, which would be difficult to encode using only eight bits. It is also important to note that the CIE 1976 Uniform Chromacity Scales are only approximately perceptually uniform, and in fact the 8-bit encoding given in Eq. (5.4) may lead to just visible quantization errors, especially for blue and pink hues. However, such artifacts should be hardly noticeable in complex images.

The LogLuv encoding has a variant which uses only 24 bits per pixel and still offers sufficient precision. However, this format can be ineffective to compress using arithmetic coding, due to discontinuities resulting from encoding two chrominance channels with a single lookup value.

5.1.4   RGB Scale: Low-Complexity RGBE Coding
The RGB Scale or the RGBS encoding simplifies the RGBE format (Section 5.1.2) to avoid expensive exponential functions:

$$(r, g, b) = (R, G, B) \cdot 16S. \qquad (5.5)$$

The encoding was used in Valve's game engine to store HDR textures and buffers using 8-bit RGBA (three channels + alpha buffer) textures [43]. The disadvantage of this approach is a limited dynamic range of about 6 $\log_{10}$ units.

5.1.5   LogYuv: Low-Complexity LogLuv

For the applications, where the complexity of the CIE 1976 Uniform Chromacity Scales is not acceptable, a simplified version of the LogLuv encoding (Section 5.1.3) can be used:

$$(\overline{Y}, \overline{u}, \overline{v}) = \left(\log_2 Y, w_b \frac{b}{Y}, w_r \frac{r}{Y} \right), \tag{5.6}$$

where Y is the luminance term computed as

$$Y = w_r r + w_g g + w_b b, \tag{5.7}$$

and the constants are equal $w_r = 0.299$, $w_g = 0.587$, $w_b = 0.114$. With nonzero and positive input r, g, and b values in the range from 2^{-16} to 2^{16}, the log-luminance $\overline{Y}$ is in the range $[-16,16]$, and the chroma components are in the range $[0,1]$ with $\overline{u} + \overline{v} \leq 1$. Unlike LogLuv, this simplified encoding cannot be used to store color values outside the red–green–blue color triangle given by the primaries. Such encoding was used for high dynamic range texture compression [44].

5.1.6   JND Steps: Perceptually Uniform Encoding

Most of the low dynamic range image or video formats use so-called gamma correction to convert luminance or RGB spectral color intensity into integer numbers, which can be latter encoded. Gamma correction is usually given in a form of the power function *intensity = signal$^\gamma$* (or *signal = intensity$^{(1/\gamma)}$* for an inverse gamma correction), where the value of γ is between (1.8) and (2.2). Gamma correction was originally intended to reduce camera noise and to control the current of the electron beam in CRT monitors (for details on gamma correction, see [45]). Accidentally, light *intensity* values, after being converted into *signal* using the inverse gamma correction formula, correspond usually well with our perception of lightness. Therefore, such values are also well suited for image encoding since the distortions caused by image compression are equally distributed across the whole scale of *signal* values. In other words, altering *signal* by the same amount for both small values and large values of signal should result in the same magnitude of visible changes. Unfortunately, this is only true for a limited range of luminance values, in practice up to 100 cd/m^2. This is because the response characteristics of the human visual system (HVS) to luminance[2] changes considerably above

---

[2]HVS use both types of photoreceptors, cones and rods, in the range of luminance approximately from 0.01 to 10 cd/m^2. Above 100 cd/m^2 only cones contribute to the visual response.

FIGURE 5.5: 28-bit per pixel JND encoding.

100 cd/m². This is especially noticeable for HDR images, which can span the luminance range from 10^{-5} to 10^{10} cd/m². An ordinary gamma correction is not sufficient in such case and a more elaborate model of luminance perception is needed. This problem is solved by the *JND* encoding, described below.

The *JND* encoding is a further improvement over the *LogLuv* encoding (see Section 5.1.3), which takes into account more accurate characteristic of the human eye. The *JND* encoding can also be regarded as an extension of gamma correction to HDR pixel values. The name *JND* encoding is motivated by its design, which makes the encoded values correlate with the just noticeable differences (JND) of luminance.

The *JND* encoding requires two bytes to represent color and 12 bits to encode luminance (see Fig. 5.5). Similar to *LogLuv* encoding, chroma is represented using the u' and v' chromaticities as recommended by CIE 1976 Uniform Chromacity Scales (UCS) diagram. Luma, l, is found from absolute luminance values, y [cd/m²], using the following formula:

$$l_{\text{hdr}}(y) = \begin{cases} a \cdot y & \text{if } y < y_l \\ b \cdot y^c + d & \text{if } y_l \le y < y_h \\ e \cdot \log(y) + f & \text{if } y \ge y_h. \end{cases} \quad (5.8)$$

There is also a formula for the inverse conversion, from 12-bit luma to luminance:

$$y(l_{\text{hdr}}) = \begin{cases} a' \cdot l_{\text{hdr}} & \text{if } l_{\text{hdr}} < l_l \\ b'(l_{\text{hdr}} + d')^{c'} & \text{if } l_l \le l_{\text{hdr}} < l_h \\ e' \cdot \exp(f' \cdot l_{\text{hdr}}) & \text{if } l_{\text{hdr}} \ge l_h. \end{cases} \quad (5.9)$$

The constants are given in the following table.

$a = 17.554$	$e = 209.16$	$a' = 0.056968$	$e' = 32.994$
$b = 826.81$	$f = -731.28$	$b' = 7.3014e - 30$	$f' = 0.0047811$
$c = 0.10013$	$y_l = 5.6046$	$c' = 9.9872$	$l_l = 98.381$
$d = -884.17$	$y_h = 10469$	$d' = 884.17$	$l_h = 1204.7$

The above formulas have been derived from the luminance detection thresholds is such a way that the same difference of values l, regardless whether in a bright or in a dark region, corresponds to the same visible difference[3]. Neither luminance nor the logarithm of luminance has this property, since the response of the human visual system to luminance is complex and nonlinear. The values of l lay in the range from 0 to 4095 (12 bit integer) for the corresponding luminance values from 10^{-5} to 10^{10} cd/m^2, which is the range of luminance that the human eye can effectively see (although the values above 10^6 would mostly be useful for representing the luminance of bright light sources). If desired, the values of l can be rescaled to lower range, in order to encode luminance using 10 or 11 bits. Such lower bit encodings should still offer quantization errors below the visibility thresholds, especially for video encoding.

A useful property of the function given in Eq. (5.8) is that it is smooth (C^1-continuous) and defined for the full positive range of luminance values, including the point $y = 0$, in which $l = 0$.

Function $l(y)$ (Eq. (5.8)) is plotted in Fig. 5.6 and labeled *JND* encoding. Note that both formula and shape of the *JND* encoding are very similar to the nonlinearity (transfer function) used in the sRGB color space [48]. Both the *JND* encoding and the sRGB nonlinearity follow similar curve on the plot, but the *JND* encoding is more conservative (a steeper curve means that a luminance range is projected on a larger number of discrete luma values, V, thus lowering quantization errors). sRGB nonlinearity consists of two segments: a linear and a power function. So does the *JND* encoding, but it additionally includes a logarithmic segment for the luminance values greater than 1420.7 (see Eq. (5.8)).

For comparison, Fig. 5.6 also shows the *log* luminance encoding, used in the *LogLuv* TIFF format. The shape of the logarithmic function is significantly different from both the sRGB nonlinearity and the *JND* encoding. Although the logarithmic function is a simple and often used approximation of the HVS response to the full range of luminance, which adheres to the Weber–Fechner law, it is clear that such approximation is very coarse and does not predict the loss of sensitivity for the low light conditions.

One difficulty that arises from the JND luminance encoding is that the luminance must be given in absolute units of cd/m^2. This is necessary since the performance of the HVS is affected by the absolute luminance levels and the contrast detection thresholds are significantly higher for low light conditions. The major source of this problem is the existing HDR capture techniques, such as multi-exposure methods, which give a measurement of relative luminance (luminance factor), but give no information on absolute luminance levels. The conversion from relative to absolute luminance units is however very simple and requires multiplication of all XYZ color coordinates by a single constant. Such a constant needs to be measured only once for a camera. The measurement can be done by capturing a scene containing a uniform light source of known luminance or a surface of measured luminance [29]. If such a measurement

---

[3]Derivation of this function can be found in [46]. The formulas are derived from the threshold versus intensity characteristic measured for human subjects and fitted to the analytical model [47].

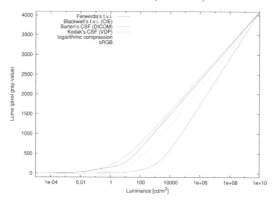

FIGURE 5.6: Functions mapping physical luminance y to encoded luma values l. JND Encoding—perceptual encoding of luminance; sRGB—nonlinearity (gamma correction) used for the sRGB color space; logarithmic compression—logarithm of luminance, rescaled to 12-bit integer range. Note that encoding high luminance values using the sRGB nonlinearity (dashed line) would require significantly larger number of bits than the perceptual encoding.

is not possible, an approximate calibration of an image to absolute units, by assuming typical luminance levels of some objects (e.g., the sky or a daylight illuminated wall), is usually sufficient.

The maximum quantization errors for all luminance encodings described in this chapter are shown in Fig. 5.7. All but the *JND* encoding have approximately uniform maximum

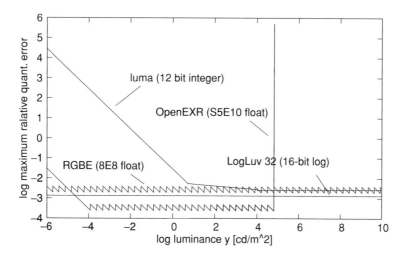

FIGURE 5.7: Comparison of the maximum quantization errors for different luminance to luma encodings: JND encoding (12-bit integer) is given by Eq. (5.8); RGBE is an encoding used in the Radiance HDR format; 16-bit half is a 16-bit floating point format used in OpenEXR; 32-bit LogLuv is a logarithmic luminance encoding used in LogLuv TIFF format.

quantization error across all visible luminance values. The edgy shape of both RGBE and *16-bit half* encodings is caused by rounding of the mantissa. The *JND* encoding varies the maximum quantization error across the range to mimic loss of sensitivity in the HVS for low light levels. This not only makes better use of the available range of luma values, but also reduces invisible noise in very dark scenes, which would otherwise be encoded. Such noise reduction can significantly improve image or video compression.

5.2   HIGH FIDELITY IMAGE FORMATS

The need for image formats capable of encoding higher dynamic ranges was recognized very early in several fields, such as computer graphics, medical imaging, or film scanning in the motion picture production. These led to several image formats, which can be classified into three following groups.

- Formats originally designed for high dynamic range images. The quantities they store are usually floating point values of a linear radiance or luminance factor[4]. There are several high-precision formats, such as Radiance's RGBE, logLuv TIFF, and OpenEXR. These formats are lossless up to the precision of their pixel representation.

- Formats designed to store as many bits as a particular sensor can provide, for example 12-bit for a film scanner. This group includes Digital Picture Exchange *DPX* format used in the movie industry to store scanned negatives, *DICOM* format for medical images, and a variety of so-called *RAW* formats used in digital cameras. All these formats use more than eight bits to store luminance, but they are usually not capable of storing such an extended dynamic range as the HDR formats.

- Formats that store larger number of bits but are not necessary intended for HDR images. Twelve or more bits can be stored in JPEG-2000 and TIFF files. All these formats can easily encode HDR if they take advantage of a pixel encoding that can represent full visible range of luminance and color gamut, such as those described in Section 5.1.

Variety of formats and lack of standards hinders the transition from traditional output-referred LDR formats to scene-referred HDR formats. The HDR formats (Radiance's RGBE, logLuv TIFF, and OpenEXR) have not gained widespread acceptance mainly because they offer only lossless compression resulting in huge files sizes. The most successful OpenEXR format has been however integrated with several Open Source and commercial applications, such as Adobe® Photoshop® starting from the release CS2. Other specialized formats, such

---

[4]For the explanation of luminance factor, refer to Section 2.1.

as DPX, DICOM, and cameras' RAW formats, usually do not allow storing as high dynamic range as the HDR formats. Since they are designed to be used for a specific application, it is unlikely that they will evolve into general purpose image formats.

The following subsections describe the two most popular HDR image formats: the Radiance HDR and the OpenEXR format.

5.2.1   Radiance's HDR Format

One of the first HDR image formats, which gained much popularity, was introduced in 1989 into the Radiance rendering package[5]. Therefore, it is known as the Radiance picture format and can be recognized by the file extensions .hdr or .pic. The file consists of a short text header, followed by run-length encoded pixels. Pixels are encoded using the XYZE or RGBE pixel formats, discussed in Section 5.1.2. The difference between both formats is that the RGBE format uses red, green, and blue primaries, while the XYZE format uses the CIE 1931 XYZ primaries. As a result, the XYZE format can encode the full visible color gamut, while the RGBE is limited to the chromaticities that lie within the triangle formed by the red, green, and blue color primaries. For more details on this format, the reader should refer to [49] and [6, Sec. 3.3.1].

5.2.2   OpenEXR

The OpenEXR format or (the Extended Range format), recognized by the file name extension .exr, was made available with an open source C++ library in 2002 by Industrial Light and Magic (see http://www.openexr.org/ and [50]). Before that date the format was used internally by Industrial Light and Magic for the purpose of special effect production. The format is currently promoted as a special-effect industry standard and many software packages already support it. Some features of this format include the following.

- Support for 16-bit floating-point, 32-bit floating-point, and 32-bit integer pixels.

- Multiple lossless image compression algorithms. Some of the included codecs can achieve 2:1 lossless compression ratios on images with film grain.

- *Extensibility.* New compression codecs and image types can easily be added by extending the C++ classes included in the OpenEXR software distribution. New image attributes (strings, vectors, integers, etc.) can be added to OpenEXR image headers without affecting backward-compatibility with existing OpenEXR applications.

---

[5]Radiance is an open source light simulation and realistic rendering package. Home page: http://radsite.lbl.gov/radiance/.

Although the OpenEXR file format offers several data types to encode channels, color data are usually encoded with 16-bit floating point numbers, known as half-precision floating point, discussed in Section 5.1.1.

5.3   HIGH FIDELITY VIDEO FORMATS

The developments in the display and digital projection technologies motivated work on high fidelity video formats. This section reviews recent advancements in this area.

5.3.1   Digital Motion Picture Production

Digital motion picture production involves processing higher dynamic range images than normally found in standard imaging setups. Cinematographic cameras capture the dynamic range up to 12 f-stops and the films are scanned to the 12-bit logarithmic DPX format. Computer-generated sequences are rendered in linear-luminance units and stored using HDR file formats.

To standardize the formats used to exchange materials involved in the motion picture production, the Science and Technology Council of the Academy of Motion Picture Arts and Sciences formed an Image Interchange Framework committee. The committee is to define a conceptual framework, file formats, and recommended practices related to color management and exchange of digital images during motion picture production and archiving. As of 2007, the standardization is an ongoing process and an early overview of the proposal can be found in [51]. The proposed framework employs the OpenEXR HDR image format (refer to Section 5.2.2) for storage and the Color Transformation Language for color profiling. Since this is going to be the first device independent framework, which does not rely on output-referred formats, this section mentions its major concepts.

The image interchange framework assumes that all original material, including scanned film negatives, images from digital cameras, and 3D computer graphics are imported into a common pixel format called "Academy Color Encoding Space" or ACES. The ACES assumes unlimited color gamut and dynamic range. It is neither output-referred, nor strictly scene-referred representation. It assumes that pixel values are approximately linear to radiance and luminance (as for most HDR file formats), but it does not require that these values correspond to the actual physical color values found in the original scenes. This is dictated by the common practices of film making, where the colors of the original scene are intentionally altered. To display ACES images, two color transform needs to be applied: the *rendering transform* gives desired "look," while the *output device transform* accounts for differences between output devices, such as preview monitors or film printers. Image editing and compositing visual effects is performed on the ACES images, stored in OpenEXR files. All color transforms are specified using the Color Transformation Language (CTL).

While the framework is still under development, it introduces several appealing concepts. The output device transform eliminates the dependence on the output device. The rendering transform introduces a flexible "tone-mapping" step, which can be altered to change the desired "look" of images. Finally, the ACES file format ensures that no information is lost due to gamut clamping or insufficient precision of the pixel format.

5.3.2   Digital Cinema

High fidelity image formats are required not only in the motion picture production process, but also when the final version of a movie is distributed and shown in movie theaters. Analog movie projectors, which still offer outstanding resolution and larger dynamic range than traditional displays, are being replaced with their digital counterparts, mostly because of much lower costs of movie distribution. The quality of digital projection is found to be comparable with the highest quality analog projection, but does not require expensive process of printing thousands of film copies.

A consortium of movie studios formed Digital Cinema Initiatives or DCI with a goal to establish a standard framework for digital movie distribution and projection. In 2007, the DCI released an updated version of the specification (v1.1). The specification assumes that single frames are encoded at the resolution 2048×1080 (2K) at 24 Hz or 48 Hz, or 4096×2160 (4K) at 24 Hz using JPEG2000. Pixels are represented as the CIE 1931 XYZ absolute trichromatic color values, so that Y value corresponds to luminance. Each trichromatic color value is normalized by the constant 52.37, compressed with the power function of the $\gamma = 2.6$ and encoded on 12-bit. The value 52.37 is slightly higher than the peak luminance of a typical projector and sets the upper threshold on the luminance that can be represented.

The specification takes great care of color data handling and making sure that the experience of digital cinema does not differ much from analog projection. This is manifested in quite moderate frame-rate of 24 Hz, which is typical to an analog film. This assumption however, and in particular the choice of the peak luminance of 52.37 cd/m^2 and the steep gamma function, makes the framework less suitable for high dynamic range movie projection.

5.3.3   MPEG for High-Quality Content

The need for encoding high fidelity video has been recently the focus of the Joint Video Team (JVT), which works on the family of popular MPEG standards. The JVT has recently added five new profiles intended for high-quality content to the MPEG4-AVC/H.264 video coding standard [52]. The new profiles offer chroma channels encoding without subsampling and with the same precision as the luma channel, so-called 4:4:4 video format coding, bit depths up to 14 bits per sample and a set of supplemental enhancement information (SEI) messages

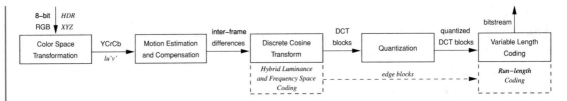

FIGURE 5.8: Simplified pipeline for the standard MPEG video encoding (black, solid) and proposed extensions (italic, dashed) for encoding High Dynamic Range video. Note that *edge blocks* are encoded together with DCT data in the HDR flow.

that describes the tone-mapping curve to map higher bit-depth content to lower number of bits.

The new profiles offer possibility of using the extended gamut color spaces, defined by IEC 61966-2-4 ($xvYCC_{601}$ and $xvYCC_{709}$) and ITU-R BT.1361. These color spaces are also optional encodings for the high-definition multimedia interface (HDMI v1.3). They can encode highly saturated colors, while maintaining backward-compatibility with the color spaces used for video coding (BT.601 and BT.709). This was possible since both BT.601 and BT.709 recommended using only the values within the range from 16 to 235, thus allowing for undershoot and overshoot found in analog TV signaling. Since such code-value margins are not necessary for digital video, they can be used to encode extended color gamut. Unfortunately, the new color space extends color gamut only toward more saturated colors, while offering the same dynamic range as the BT.601 and BT.709; therefore, it is not suitable for encoding HDR content.

5.3.4   HDR Extension of MPEG-4

It was demonstrated [53] that the MPEG encoding standard, both the Advanced Simple Profile (ISO/IEC 14496-2) [54] and the Advanced Video Coding (ISO/IEC 14496-10) [55], can be extended to handle HDR data. The scope of required changes to MPEG-4 encoding is surprisingly modest. Figure 5.8 shows a simplified pipeline of MPEG-4 encoding, together with proposed extensions. While a standard MPEG-4 encoder takes as input three 8-bit RGB color channels, the HDR encoder must be provided with pixel values in the absolute XYZ color space [7]. Such color space can represent the full color gamut and the complete range of luminance the eye can adapt to. Next pixel values are transformed to the color space that improves the efficiency of encoding. MPEG-4 converts pixel values to one of the family of YC_BC_R color spaces, which exhibit low correlation between color channels for natural images. The proposed extension uses instead the perceptually uniform HDR pixel encoding, described in Section 5.1.6. The 11-bit, instead of 12-bit, encoding of luma is used as it turns out to be both conservative and easy to introduce to the existing MPEG-4 architecture.

Due to quantization of DCT coefficients, noisy artifacts may appear near edges of high-contrast objects. This problem is especially apparent for HDR video, in particular for synthetic sequences, where the contrast tends to be higher than in natural LDR video. This can be alleviated by encoding sharp-contrast edges in each 8×8 block separately from the rest of the signal. An algorithm for such hybrid encoding can be found in [53].

Additional examples and the demonstration video can be found on the project web page: `http://www.mpi-inf.mpg.de/resources/hdrvideo/index.html`.

5.4 BACKWARD-COMPATIBLE COMPRESSION

Since the standard low dynamic range (LDR) file formats for images and video, such as JPEG or MPEG, have become widely adapted standards supported by almost all software and hardware equipment dealing with digital imaging, it cannot be expected that these formats will be immediately replaced with their HDR counterparts. To facilitate transition from the traditional to HDR imaging, there is a need for backward-compatible HDR formats, that would be fully compatible with existing LDR formats and at the same time would support enhanced dynamic range and color gamut. Moreover, if such a format is to be successful and adopted by large part of the market, the overhead of HDR information must be very low, preferably below 30% of the LDR file size. This is because very few consumers will have access to HDR technology, such as HDR displays, at the beginning and the rest of the consumers will not accept doubling the size of the file for the sake of the data they cannot take advantage of. Such backward-compatible encoding would also require that the original LDR content is not modified. Although the compression of HDR can be improved if an LDR image can be slightly altered, this would also be unacceptable for majority of applications where it is crucial to preserve the original appearance of LDR content.

The following subsections present an overview of both existing and potential solutions for backward-compatible image and video encoding.

5.4.1 JPEG HDR

Spaulding et al. [56] showed that the dynamic range and color gamut of typical sRGB images can be extended using residual images. Their method is backward-compatible with the JPEG standard, but only considers images of moderate dynamic range. Ward and Simmons [57] have proposed a backward-compatible extension of JPEG, which enables compression of images of much higher dynamic range (JPEG HDR). JPEG HDR is the extension of the JPEG format for storing HDR images that is backward-compatible with an ordinary 8-bit JPEG. A JPEG HDR file contains a tone-mapped version of an HDR image and additionally a ratio (subband)

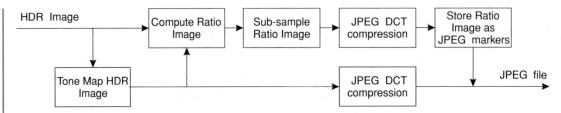

FIGURE 5.9: Data flow of subband encoding in JPEG HDR format.

image, which contains information needed to restore the HDR image from the tone-mapped image. The ratio image is stored in the user-data space of JPEG markers, which are normally ignored by applications. This way, a naive application will always open the tone-mapped version of an image, whereas an HDR-aware application can retrieve the HDR image.

A data flow of the subband encoding is shown in Fig. 5.9. An HDR image is first tone mapped and compressed as an ordinary JPEG file. The same image is also used to compute the ratio image, which stores a ratio between HDR and tone-mapped image luminance for each pixel. To improve encoding efficiency, the ratio image is subsampled and encoded at lower resolution using the ordinary JPEG compression. The compressed sub-band image is stored in the JPEG markers. To reduce the loss of information due to subsampling the ratio image, two correction methods have been proposed: enhancing edges in a tone-mapped image (so-called *pre-correction*) and synthesizing high frequencies in the ratio image during up-sampling (so-called *post-correction*). Further details on the JPEG HDR compression can be found in [57] and [58].

5.4.2 Wavelet Compander

Li et al. [59] propose that HDR images can be encoded using only 8-bits, if they undergo a reversible companding operation. They propose a multiscale wavelet architecture, which can compress an HDR image to a lower bit-depth and later expand it to obtain a result that is close to the original HDR image. The information loss is reduced by amplifying low amplitudes and high frequencies at the compression stage, so that they survive the quantization step to the 8-bit LDR image. Such technique is conceptually similar to the *pre-correction* in JPEG HDR. Since the expansion is a fully symmetric inverted process, the amplified signals are properly suppressed to their initial level in the companded HDR image. To further reduce the information loss, the compressed image is iteratively modified to improve the correlation of its subbands with respect to the original HDR image. The authors observe a good visual quality of both the compressed and companded images, but they admit that any guarantee concerning their fidelity to tone mapped (i.e., undergoing just one compression iteration) and original HDR images cannot be given. The obtained PNSR for the companded HDR image is even worse than for ordinary LUT (Look-Up-Table) companding; however, the results of the multi-scale wavelet companding look visually better.

Given the requirements for a backward-compatible image and video compression, the lack of fidelity of tone-mapped images is often not acceptable, since the original material quality cannot be compromised. Another limitation of this technique is fixed tone-mapping operator. The emphasis on high frequencies at the compression step makes the proposed framework less suitable for standard JPEG and MPEG techniques, which use the quantization matrices that are perceptually tuned to discard visually non-important high frequencies. This is confirmed by relatively poor compression rates reported the authors when they attempted to combine JPEG with their companding. It is not clear, how the compander approach can be adopted for lossy HDR video compression, in which temporal coherence and computation efficiency must be guaranteed.

5.4.3   Backward-Compatible HDR MPEG

Encoding of movies in high fidelity format is becoming more important as the quality of consumer-level displays is starting to exceed the quality of available DVD or broadcast content. As discussed in Section 5.3.1, high fidelity content is available at the movie production stage. However, to encode motion pictures using traditional MPEG compression, the movie must undergo processing called color grading. Part of this process is the adjustment of tones (tone mapping) and colors (gamut mapping), so that they can be displayed on majority of TV sets (refer to Fig. 5.10). Although such processing can produce high quality content for typical CRT and LCD displays, the high-quality information, from which advanced HDR displays could benefit, is lost.

The HDR-MPEG encoding, similarly as the JPEG-HDR (refer to Section 5.4.1), compresses both LDR and HDR video stream and stores them in the same backward-compatible movie file (see Fig. 5.10). Depending on the capabilities of the display and playback hardware

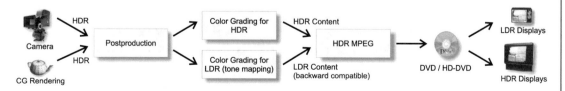

FIGURE 5.10:   The proposed backward-compatible HDR DVD movie-processing pipeline. The high dynamic range content, provided by advanced cameras and CG rendering, is encoded in addition to the low dynamic range (LDR) content in the video stream. The files compressed with the proposed HDR MPEG method can play on existing and future HDR displays.

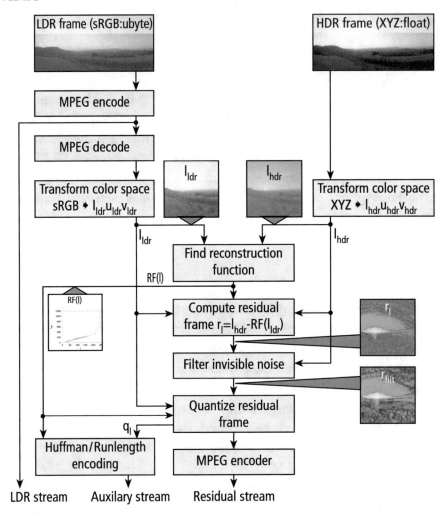

FIGURE 5.11: A data flow of the backward-compatible HDR MPEG encoding. See text for details.

or software, either LDR or HDR content is displayed. This way HDR content can be added to the video stream at the moderate cost of about 30% of the LDR stream size.

The complete data flow of the HDR-MPEG compression algorithm is shown in Fig. 5.11. The encoder takes two sequences of HDR and LDR frames as input. The LDR frames, intended for LDR devices, usually contain a tone-mapped or gamut-mapped version of the original HDR sequence. The LDR frames are compressed using a standard MPEG encoder (*MPEG encode* in Fig. 5.11) to produce a backward-compatible LDR stream. The LDR frames are then decoded to obtain a distorted (due to lossy compression) LDR sequence, which is later used as a reference for the HDR frames (see *MPEG decode* in Fig. 5.11).

Both the LDR and HDR frames are then converted to compatible color spaces, which minimize differences between LDR and HDR colors. For the HDR pixels, the JND encoding discussed in Section 5.1.6 is used. For the LDR pixels, the CIE 1976 Uniform Chromacity Scales (Eqs. (5.3) and (5.4)) are used for chrominance and the sRGB nonlinear transfer function is used to encode luminance.

The reconstruction function (see *Find reconstruction function* in Fig. 5.11) reduces the correlation between LDR and HDR pixels by giving the best prediction of HDR pixels based on the values of LDR pixels. The residual frame is introduced to store a difference between the original HDR values and the values predicted by the reconstruction function.

To improve compression, invisible luminance, and chrominance variations are removed from the residual frame (see *Filter invisible noise* in Fig. 5.11). Such filtering simulates the visual processing that is performed by the retina to predict the contrast detection threshold at which the eye does not see any differences. The contrast magnitudes that are below this threshold are set to zero. An example of such filtering is shown in Fig. 5.12.

Finally, the pixel values of the residual frame are quantized (see *Quantize residual frame* in Fig. 5.11) and compressed using a standard MPEG encoder into a residual stream. Both the

FIGURE 5.12: Residual frame before (left) and after (center) filtering invisible noise. Such filtering removes invisible information, while leaving important high-frequency details that are lost if ordinary low-pass filtering (downsampling) is used (right). Green color denotes negative and gray positive values. The *Memorial Church* image courtesy of Paul Debevec.

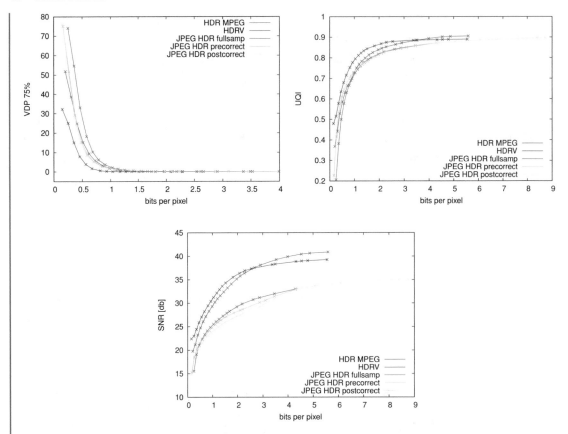

FIGURE 5.13: Comparison of lossy HDR compression algorithms. Metrics: VDP 75%—HDR-VDP percentage of visibly different pixels at $P = 75\%$; UQI—universal quality index [60]; SNR—signal-to-noise-ratio for the JND-encoded luma (refer to Section 5.1.6). The results are averaged for a set of images.

reconstruction function and the quantization factors are compressed using a lossless arithmetic encoding and stored in an auxiliary stream.

The encoding scheme was tested with a number of tone-mapping operators, with and without invisible noise filtering step, and compared to other HDR compression methods. The best performance was achieved for global tone-mapping operators, which do not amplify high frequencies. As shown in Fig. 5.13, the HDR-MPEG compression performed worse than the HDR extension of MPEG-4 (refer to Section 5.3.4), labeled as *HDRV*. This is because the HDRV encoding is not backward-compatible and therefore does not need to encode any information on an LDR stream. For the HDR VDP and the UQI metrics, the JPEG HDR (refer to Section 5.4.1) performs almost the same as the HDR MPEG for the pre-correction and the post-correction approach, but is worse for the full-sampling, even though the HDR MPEG does not involve subsampling. JPEG HDR performs slightly worse than the HDR-MPEG for the SNR metric.

More information on this project as well as the demonstration video can be found on the project web page: http://www.mpii.mpg.de/resources/hdr/hdrmpeg/.

5.4.4 Scalable High Dynamic Range Video Coding from the JVT

The JVT, responsible for the family of MPEG standards, considers several proposals for the scalable bit-depth coding. The scalable bit-depth coding is equivalent to the backward-compatible coding (hence scalability) that can store HDR data (hence enhanced bit-depth). This naming convention is taken from the spatial scalable coding that provides higher resolution and the temporal scalable coding that offers higher frame-rate. The proposed extensions are conceptually similar to the JPEG HDR introduced in Section 5.4.1 and the HDR MPEG described in Section 5.4.3. They encode a tone-mapped sequence using a backward-compatible 8-bit coding, a series of coefficient for predicting HDR frames based on tone-mapped frames (inter-layer prediction), and a residual stream that encodes prediction errors. In contrast to JPEG HDR and HDR MPEG, the proposed schemes focus on computational efficiency; therefore, they use simplified color transforms and avoid expensive arithmetic operations.

One of the proposals [61] suggests using the following transform to predict the high dynamic range pixel chroma and luma components based on a tone-mapped pixel value:

$$Y_{\mathrm{HDR}} = \alpha\ Y_{\mathrm{LDR}} + \mathrm{offset},$$

$$Cb_{\mathrm{HDR}} = \alpha\ Cb_{\mathrm{LDR}} + \mathrm{offset} \cdot \frac{Cb_{\mathrm{LDR,DC}}}{Y_{\mathrm{LDR,DC}}},$$

$$Cr_{\mathrm{HDR}} = \alpha\ Cr_{\mathrm{LDR}} + \mathrm{offset} \cdot \frac{Cr_{\mathrm{LDR,DC}}}{Y_{\mathrm{LDR,DC}}}$$

(5.10)

where the α and "offset" are prediction coefficients stored for each block and $Y_{\mathrm{LDR,DC}}$, $Cb_{\mathrm{LDR,DC}}$, $Cr_{\mathrm{LDR,DC}}$ are the DC portion (mean) of the luma and chroma components in the LDR image block. The non-intuitive part of this transform is the presence of luma component in the prediction of Cb_{HDR} and Cr_{HDR}. Such normalizing luma factor is necessary, since most color spaces utilized for video coding are not iso-luminant, which means that they contain some luma information in their chroma components. The division by the $Y_{\mathrm{LDR,DC}}$ reduces the variance in chroma due to luma component. The HDR MPEG coding solves this problem by employing approximately iso-luminant CIE 1976 Uniform Chromacity Scales (Eqs. (5.3) and (5.4)).

5.5 HIGH DYNAMIC RANGE TEXTURE COMPRESSION

High dynamic range textures can significantly enhance realism in real-time rendering using graphics hardware. This is, however, achieved at the cost of higher memory footprint, which can affect rendering performance. The bottleneck is both graphics card memory and the bandwidth available for sending textures from external to graphics card memory. Both these problems can be reduced if textures are efficiently compressed prior to sending them to graphics card memory.

There exist several common techniques for compressing low dynamic range textures. The S3TC texture compression scheme, also known as DXTC [62], has became a de facto standard that is often implemented on graphics cards. It divides a texture into 4×4 blocks, then encodes each block using 64 bits, resulting in 4 bits per pixel. Two colors are selected as base colors and stored in 16 bits (red—5 bits, green—6 bits, blue—5 bits). Then each pixel is encoded in 2 bits, which are used to linearly interpolate between the two base colors. Unlike video or image compression, texture compression schemes are always fixed-rate to allow random access to texels. They must be also simple enough to offer very fast decoding and to be suitable for hardware implementation. Unfortunately, a straightforward extension of the S3TC to larger number of bits that could encode HDR textures results in visible quantization and blocking artifacts [63]; therefore, more elaborate compression schemes are necessary.

Munkberg et al. [44] extended the S3TC scheme for high dynamic range luminance data and proposed an interesting approach to chroma encoding. The pixels are initially transformed to the $\log Y\overline{u}\overline{v}$ color space, described in Section 5.1.5. The luminance is coded similarly as in the S3TC scheme, using two base log-luminance values encoded in 8 bits and 16 (for a 4×4 block) 4-bit indexes used to interpolate between the base values. The interpolation can be both uniform or non-uniform with smaller steps close to the base values. The chroma channel is subsampled either horizontally or vertically, halving the number of pixels to encode. Two base chroma colors are coded in 15 bits each (8 bits for $\overline{u}$ and 7 bits for $\overline{v}$) together with eight

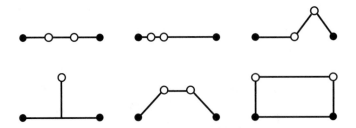

FIGURE 5.14: Shapes used for coding chroma in a 4×4 texture block. Redrawn from [44].

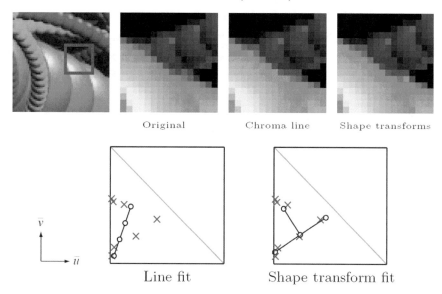

FIGURE 5.15: Shape transforms can fit better to color distribution in a block than linear interpolation. Images courtesy of Jacob Munkberg, Petrik Clarberg, Jon Hasselgren, and Tomas Akenine-Möller of the Lund University [44]. © 2006 ACM, Inc. Reprinted by permission.

2-bit indexes. Then instead of using a straight line for interpolating between the two base colors, Munkberg et al. suggest to use one of the predefined shapes, shown in Fig. 5.14. The two base colors are used to fix the position of two vertices (solid vertices in the figure), thus allowing for shifting, scaling, and rotation of the predefined shapes in the $\overline{uv}$ coordinates. Then each chroma index indicates the vertex that should be used to decode chroma. An example illustrating the difference between linear interpolation and shape transform coding is shown in Fig. 5.15. In overall, the Munkberg's et al. compression scheme requires 128 bits per 4×4 block, thus 8 bits per pixel, instead of 48 bits required for the half precision floating point buffers (see Section 5.1.1).

Roimela et al. [63] propose to abandon the S3TC scheme and use the properties of the floating point numbers in their HDR texture compression method. Similarly as Munkberg's et al. scheme, the proposed encoding operates on 4×4 blocks, each encoded in 128 bits or 8 bits per pixel. The first 72 bits are used to encode luminance. The luminance of each pixel is encoded separately using 4 bits, with a common exponent (5 bits) and the number of leading zeros (3 bits) shared for each pixel in a block. To compute chroma pixels, red and blue color components are divided by luminance values. Then, the chroma pixels are subsampled both horizontally and vertically, reducing the number of color samples to 4. Each color component is quantized into seven bits, so that 4 color samples × 2 color components × 7 bits can be encoded into the remaining 56 bits. Roimela's et al. texture compression scheme locally adapts to the dynamic range of a block, resulting in coarser quantization for high contrast blocks and lower quantization for the blocks in smooth regions. Another useful property is relatively low complexity and decoding to the half-precision floating point numbers, which are supported by graphics cards.

Both Munkberg's et al. and Roimela's et al. texture compression schemes require modifications of the graphics hardware for the best performance. Therefore, it can be expected that future work on HDR texture compression will focus on the schemes that allow for efficient decoding on existing hardware using fragment programs.

5.6   CONCLUSIONS

It is quite surprising that the well-studied and improved over years general image and video compression standards may turn out to be inadequate for new content and displays in the coming years. Although increasing the bit-depth of encoded images seems to be the most apparent solution to this problem, it does not address the major issue: how the encoded code-values should be mapped to the luminance levels produced by a display. The ICC color profiles, commonly used for this purpose in low dynamic range images, have been designed for reflective print colorimetry and are not suitable for high contrast displays. The problem is even more difficult if the output device is unknown and may vary from a low-contrast mobile display to a high-end large screen display. To fully address this issue, not only the compression algorithms, but the entire imaging pipeline, from acquisition to display-adaptive tone-mapping, must be redesigned. High dynamic range pixel encodings (Section 5.1) offer a general purpose intermediate storage format, which can represent the colorimetrically calibrated images with no display limitations. Such images could be displayed only on an ideal display, capable of producing all physically feasible colors, which is unlikely to ever exist. Therefore, the high dynamic range images must be adjusted to the actual display capabilities by compressing its dynamic range, clipping excessively bright pixels, choosing the right brightness level, so that all colors fit into the display color gamut. The tone-mapping algorithms designed for that purpose are discussed in Chapter 6 of this book. Since making such radical changes in the imaging pipeline would render the existing software and hardware obsolete, it is important to ensure backward-compatibility of image and video formats, as discussed in Section 5.4.

The specialized application areas that require higher image and video fidelity than offered by a general purpose compression formats have already come up with custom formats, such as Radiance RGBE or OpenEXR for computer graphics animation, or DICOM for medical images, as discussed in Section 5.2. The proposals of the Image Interchange Framework committee (Section 5.3.1) work on defining not only image format, but also the entire imaging pipeline employed for digital motion picture production. Another specialized area is texture compression (Section 5.5), which have different requirements (fixed-rate coding, fast decoding) than general purpose image compression. It can be expected that some advanced ideas from these specialized compression formats will be incorporated in the future general purpose standards, such as the family of MPEG or JPEG formats.

CHAPTER 6

Tone Reproduction

The contrast and brightness range in typical HDR images exceeds capabilities of current display devices or print. Thus, these media are inadequate to directly reproduce the full range of captured light. Tone mapping is a technique for the purpose of reducing contrast and brightness in HDR images to enable their depiction on LDR devices. The process of tone mapping is performed by a tone-mapping operator.

Particular implementations of a tone-mapping operator are varied and strongly depend on a target application. A photographer, computer graphics artist, or a general user will most probably like to simply obtain nice looking images. In such cases, one most often expects a good reproduction of appearance of an original HDR scene on a display device. In simulations or predictive rendering, the goals of tone mapping may be stated more precisely: to obtain a perceptual brightness match between HDR scene and tone-mapped result, or to maintain equivalent object detection performance. In visualization or inspection applications often the most important is to preserve as much of fine detailed information in an image as possible. Such a plurality of objectives leads to a large number of different tone-mapping operators.

In this chapter, we present at first short overview of existing tone-mapping operators. Then we discuss the problem of tone-mapping evaluation using subjective and objective methods. Finally, we discuss tone-mapping extensions into temporal domain as required to handle HDR video.

6.1 TONE-MAPPING OPERATORS

Various tone-mapping operators developed in recent years can be generalized as a transfer function which takes luminance or color channels of an HDR scene as input and processes it to output pixel intensities that can be displayed on LDR devices. The input HDR image can be calibrated so that its luminance is expressed in SI units cd/m^2 or it may contain relative values which are linearly related to luminance (Section 3.2). The transfer function may be the same for all pixels in an image (global operator) or its shape may depend on the luminance of neighboring pixels (local operator). In principle, all operators reduce the dynamic range of input data. Since most of the algorithms process only luminance, color images have to be converted to a color space that decouples luminance and chrominance, e.g. Yxy (refer to Section 2.2). After tone mapping, the Yxy color space is converted to the original color space of the image. In such an inverse transform, the tone-mapped intensities are used instead of the original luminance as the Y channel, while the chrominance is left unchanged.

6.1.1   Luminance Domain Operators

The most naïve approach to tone mapping is to "window" a part of luminance range in an HDR image. That is to map a selected range of luminance using a linear transfer function to a displayable range. Such an approach, however, renders dark parts of image black and saturates light areas to white, thus removing the image details in the areas. A basic sigmoid function,

$$L = \frac{Y}{Y + 1},$$

(6.1)

maps the full range of scene luminance Y in the domain $[0, \inf)$ to displayable pixel intensities L in the range of $[0, 1)$. Such a function assures that no image areas are saturated or black, although contrast may be strongly compressed. Since the mapping in Eq. (6.1) is the same for all pixels, it is an example of a global tone-mapping operator. Other global operators include *adaptive logarithmic mapping* [64], the sigmoid function derived from photographic process: *photographic tone reproduction (global)* [65], a mapping inspired by the response of photoreceptors in the human eye: *photoreceptor* [66], a function derived through *histogram equalization* [67]. The subtle differences in tone-mapped images using these operators are illustrated in Fig. 6.1. Usually, one obtains a good contrast mapping in the medium light levels and low contrast in the dark and light areas of an image. Therefore, intuitively, the most interesting part of an image in terms of its contents should be mapped using the good contrast range. The appropriate medium brightness level for the mapping is in many cases automatically determined as a logarithmic average of luminance values in an image:

$$Y_A = \exp\left(\frac{\sum \log(Y + \epsilon)}{N}\right) - \epsilon,$$

(6.2)

where Y denotes luminance, N is the number of pixels in an image, and ϵ denotes a small constant representing the minimum luminance value to prevent 0 in logarithm. The Y_A value is then used to normalize image luminance prior to mapping with a transfer function. For example, in Eq. (6.1) such a normalization would map the luminance equal to Y_A to 0.5 intensity which is usually displayed as middle-gray (before the gamma correction). Y_A is often called the adapting luminance, because such a normalization is similar to the process of adaptation to light in human vision.

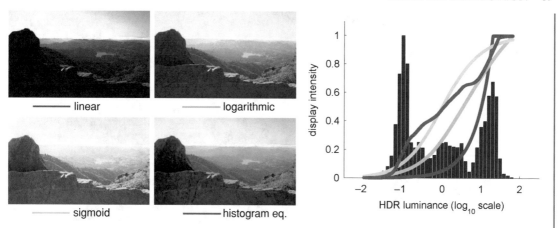

FIGURE 6.1: Comparison of global transfer functions with linear mapping (standard gamma correction with dynamic range clipping) given as the reference. The plot illustrates how luminance values are mapped to the pixel intensities on a display. The steepness of the curve determines the contrast in a selected luminance range. Luminance values for which display intensities are close to 0 or 1 are not transferred. The HDR image courtesy of Industrial Light & Magic.

6.1.2 Local Adaptation

While global transfer functions are simple and efficient methods of tone mapping, the low contrast reproduction in dark and light areas is a disadvantage. To obtain a good contrast reproduction in all areas of an image, the transfer function can be locally adjusted to a medium brightness in each area:

$$L = \frac{Y'}{Y'_L + 1},\tag{6.3}$$

where Y' denotes HDR image luminance normalized by the globally adapting luminance $Y' = Y/Y_A$ and Y'_L is the locally adapting luminance. The value of globally adapting luminance Y_A is constant for the whole image, while the locally adapting luminance Y'_L is an average luminance in a predefined area centered around each tone-mapped pixel. Practically, Y'_L is computed by convolving the normalized image luminance Y' with a Gaussian kernel. The standard deviation of the kernel σ defines the size of an area influencing the local adaptation and usually corresponds in pixels to $1°$ of visual angle. The mechanism of local adaptation is again inspired by similar processes occurring in the human eyes. Figure 6.2 illustrates the improvement in tone-mapping result through introduction of the local adaption.

The details are now well visible in dark and light areas of the image. However, along high contrast edges one can notice a strong artifact visible as dark and light outlines—the halo. The reason why such artifact appears is illustrated in Fig. 6.3. Along a high contrast edge the area of

Logarithmic average of luminance in the HDR image.

uniform global adaptation map Y_A

Gaussian blur of the HDR image with kernel size ~1deg of visual angle.

global Y_A and local Y_L' adaptation

FIGURE 6.2: Tone-mapping result with global, Eq. (6.1), and local adaptation, Eq. (6.3). The local adaptation (right) improves the reproduction of details in dark and light image areas, but introduces halo artifacts along high contrast edges.

local adaptation includes both high and low luminance; therefore, the computed average in the area is inadequate for any of them. On the side of high luminance, the local adaptation is more and more underestimated as the tone-mapped pixels are closer to the edge; therefore, Eq. (6.3) gradually computes much higher intensities than appropriate. The reverse happens on the side of low luminance. A larger blur kernel spreads the artifact over a larger area, while a smaller blur kernel reduces the artifact but also reduces the reproduction of details.

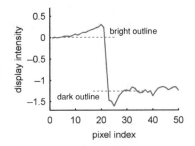
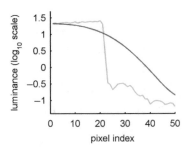

FIGURE 6.3: The halo artifact along a high contrast edge (left) and the plots illustrating the marked vertical line in tone-mapped image (middle) and HDR image (right). Gaussian blur (under-) overestimates the local adaptation (red) near a high contrast edge (green). Therefore, the tone-mapped image (blue) gets too bright (too dark) closer to such an edge.

6.1.3 Prevention of Halo Artifacts

Many image-processing techniques have been researched to prevent the halo artifacts out of which the notable solutions are automatic dodging and burning (*photographic tone reproduction (local)* [65]) and the use of *bilateral filtering* instead of Gaussian blur [68].

The automatic dodging and burning technique derives intuitively from the observation that a halo is caused by a too large adaptation area, Fig. 6.3, but also a large area is desired for a good reproduction of details. Therefore, the size of the local adaptation area is adjusted individually for each pixel location such that it is as large as possible but does not introduce halo. The halo artifact appears as soon as both very high and very low luminance values exist in an adaptation area and significantly change the estimated local adaptation. Therefore, by progressively increasing the adaptation area for each pixel, the following test can detect the appearance of halo:

$$|Y_L(x, y, \sigma_i) - Y_L(x, y, \sigma_{i+1})| < \epsilon. \tag{6.4}$$

For each pixel, the size of the adaptation area, defined by the standard deviation of the Gaussian kernel σ_i, is progressively increased until the difference between the two successive estimates is larger than a predefined threshold ϵ. The result of the Gaussian blur for the largest σ_i that passed the test is then used for given pixel in Eq. (6.3). The example of estimated adaptation areas is illustrated in Fig. 6.4. The whole process can be very efficiently implemented using the Gaussian pyramid structure as described in [65].

FIGURE 6.4: Estimated adaptation areas for pixels marked as blue cross. In each case, the green circle denotes the largest, thus the most optimal adaptation area. A slightly larger areas denoted as red circles would change the local adaptation estimate Y_L more than acceptable threshold in Eq. (6.4) and would introduce a halo artifact.

Bilateral filtering is an alternative technique to prevent halos [68]. The reason for halos, Fig. 6.3, can also be explained by the fact that the local adaptation for a pixel of high luminance is incorrectly influenced by pixels of low luminance. Therefore, excluding pixels of significantly different luminance from local adaptation estimation prevents the appearance of halo in a similar way as in Eq. (6.4). The bilateral filter [69] is a modification of the Gaussian filter which includes an appropriate penalizing function:

$$Y_L^p = \sum_{q \in N(P)} f_{\sigma_s}(\|p - q\|) \cdot Y^q \cdot g_{\sigma_r}(|Y^p - Y^q|). \qquad (6.5)$$

In the above equation, p denotes the location of the tone-mapped pixel, q denotes pixel locations in the neighborhood $N(p)$ of p. The first two terms of equation, $f_{\sigma_s} \cdot Y^q$, define Gaussian filtering with spatial σ_s. The last term, g_{σ_r}, practically excludes from the convolution those pixels whose luminance value differs from the tone-mapped one by more than σ_r. Both f and g are Gaussian functions, and luminance is usually expressed in the logarithmic space for the purpose of such filtering. The bilateral filtering process is shown in Fig. 6.5.

Compared to the automatic dodging and burning, the bilateral filter better reproduces details at the edges, because in most cases a relatively larger area is used for estimation of local adaptation. Although the exact computation of Eq. (6.5) is very expensive, a good approximation can be computed very efficiently [68, 70].

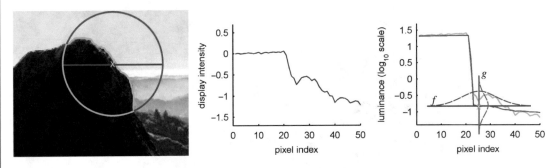

FIGURE 6.5: Bilateral filtering of a similar scanline as in Fig. 6.3, here marked in magenta (left). The penalizing function g (right plot) improves the estimation of the local adaptation (red) by excluding pixels in the neighborhood f (magenta) whose luminance value is outside the defined range (orange). Thus, the local adaptation for the pixel marked with a cross (left image) is estimated only from the pixels in the area outlined in green, while the Gaussian blur would also include pixels in the area outlined in red. The middle plot illustrates tone-mapped pixel intensities resulting from bilateral filtering.

6.1.4   Segmentation-Based Operators

An alternative approach to tone mapping, which is in a sense similar to the local adaptation techniques, is based on a fuzzy segmentation of an HDR image into areas of common and distinct illumination. Such algorithms focus on optimizing the relations of contrast or luminance between the segments while leaving the relations of pixel intensities within the segments unchanged or very simply transformed. The reduction of dynamic range can be accomplished by optimizing the whole segments because the information within a segment is usually of low dynamic range, while the differences of luminance level between the segments contribute to the high dynamic range. Unlike in local adaptation approaches which are inspired by the behavior of photoreceptors in the human eyes, the motivation here comes from the psychophysical theories of perception, mainly *Gestalt*.

One example of such an approach is the *lightness perception* tone mapping [14]. The algorithm is inspired by an anchoring theory of lightness perception [71] which comprehensively explains many characteristics of a human visual system such as lightness constancy and its spectacular failures which are important in the perception of images. The principal concept of this theory is the perception of complex scenes in terms of groups of consistent areas (frameworks). Such areas, following the Gestalt theorists, are defined by the regions of common illumination. The key aspect of the image perception is the estimation of lightness within each framework through the anchoring to the luminance perceived as white, followed by the computation of the global lightness. Lightness is a perceptual quantity that assigns brightness to the perceived shades of gray, and is judged relative to the brightness of a similarly illuminated area that appears to be white.

In such segmentation approaches, the frameworks can be identified with an automatic method for image decomposition [14] which derives from the principles of the anchoring theory of lightness perception, or alternatively by a user guidance [72]. Correspondingly, the local mapping of luminance to perceived scale of grays can be automatically adjusted with a brightness adjustment method [14, 73] or manually.

The segmentation approaches mostly do not affect the local contrast and preserve the natural colors of an HDR image due to the linear handling of luminance. The fuzzy definition of segments assures that artifacts do not appear in the areas where distinct illuminations mix. The strength of such operators is especially evident for difficult shots of real-world scenes which involve distinct regions with significantly different luminance levels, Fig. 6.6.

6.1.5   Contrast Domain Operators

The tone-mapping methods discussed so far perform the dynamic range-reducing operations directly on luminance or on color channel intensities. However, one can observe that an image with a wide range of luminance also contains a large range of contrast. Therefore,

FIGURE 6.6: The *lightness perception* tone-mapping operator reduces the contrast in HDR image (left) by decomposing the image into the areas of consistent illumination (middle) and optimizing the contrast ratio between these areas (right). In the middle image, blue and magenta illustrate the influence of two distinct frameworks and the transition between the two colors mark fuzzy areas influenced by both frameworks. The HDR image courtesy of SpheronVR.

as an alternative to luminance range compression, contrast magnitudes in the image can be reduced. Since contrast conveys semantical information in images, such a control over contrast can be advantageous. For instance, small contrast usually represents the reflectance properties of surfaces, such as texture, medium contrast often defines the outlines of objects, and large contrast represents changes in illumination. Particularly, large contrasts are in most cases the cause of a high dynamic range. By preserving small and medium contrasts, and reducing large contrasts, one can reduce the dynamic range of illumination and at the same time preserve good visibility of details from the original HDR image. Such a contrast-based processing gives a better control over transferred image information than the luminance-based operators. The latter, however, give a better control over brightness mapping. In fact, it is hard to impose a target luminance range for contrast-based compression.

A typical contrast-based tone-mapping operator includes the following steps. First, the input luminance is converted to a contrast representation. The magnitudes of contrasts are then modulated using a transfer function for contrast—the tone-mapping step. Next, the modulated contrast representation is integrated to recover the luminance information, and such luminance is then scaled to fit the available dynamic range. Finally, since the result of integration is calculated with an unknown offset, the brightness of the tone-mapped image is adjusted.

Contrast in tone-mapping applications is most often measured as a logarithmic ratio of luminance:

$$C = \log \frac{Y^p}{Y^q},$$
(6.6)

(a) HDR image, clipped   (b) contrast representation   (c) contrast transfer map   (d) tone mapping resutl

FIGURE 6.7: Contrast domain tone mapping [74]. The HDR image (a) is transformed to a contrast representation (b) which is multiplied by a contrast transfer function (c). The contrast representation is then integrated to obtain a tone-mapped image (d). In (b), white denotes strong local contrast and black no contrast. In (c), black denotes strong contrast attenuation and white marks no change in local contrast.

where Y^p and Y^q denote luminance of adjacent pixel location. The contrast representation of an image is computed as a gradient of $\log Y$, since the logarithm of division is equal to the difference of logarithms. For the tone mapping, such a representation is often multi-resolution to measure contrasts between adjacent pixels (full resolution) and adjacent areas in an HDR image (coarser resolutions). The contrasts are then modulated by a transfer function as for example in *gradient domain compression* [74]:

$$T(C) = \frac{\alpha}{|C|} \cdot \left(\frac{|C|}{\alpha} \right)^{\beta}.$$

$$(6.7)$$

Given that $\beta \in (0, 1)$, such a function attenuates gradients that are stronger than α and amplifies smaller ones. Thus, if α is equal to medium contrasts in an image, Eq. (6.7) reduces the dynamic range caused by large differences in illumination and enhances fine scale details. More complex transfer functions are also possible including for instance contrast equalization [75]. As the final step, the modulated contrast representation of an HDR image has to be integrated in order to obtain intensities in a tone-mapped image. The integration step is performed by solving the Poisson equation and the brightness adjustment step is left for manual setting by a user. The stages of the contrast domain tone mapping process are illustrated in Fig. 6.7.

6.2   TONE-MAPPING STUDIES WITH HUMAN SUBJECTS

The previous sections provide only an introduction to the general ideas behind the tone-mapping problem and the reader is referred to [6] for detailed descriptions of specific algorithms. Existing tone-mapping operators can be further generalized to a transfer function in form of a "black box" which converts scene luminance to displayable pixel intensities. While the universal goal of such a transfer function is to reduce the original dynamic range and at the same time preserve the original appearance of HDR, a particular realization of it can be versatile and depends on the objectives of a target application. In many cases, one may wish to simply obtain nice looking images that resemble the original HDRs, but the requirements may also be more precise: perceptual brightness match, good visibility of details, equivalent object detection performance in the tone-mapped and corresponding HDR image, and so on. In view of the technical limitations imposed by standard displays and other constraints related to particular image observation conditions (ambient lighting, the screen resolution, the observer distance), such requirements can only be met at the cost of other image properties. For instance, if an available dynamic range is assigned to enable good visibility of details (local contrasts), there may not be enough dynamic range left to depict global contrast variations in the scene (refer to Fig. 6.8). The tradeoff between these conflicting goals is often balanced through an

FIGURE 6.8: Different levels of detail visibility in tone-mapping results. The increase in detail visibility is obtained at the cost of contrasts between larger image areas. The image (a) is the adaptive logarithmic mapping [64], (b) is the lightness perception tone mapping [14], and (c) is the contrast domain tone mapping [75]. The HDR image courtesy of Byong Mok Oh.

optimization process, but sometimes the design of an algorithm is focused on the requirements and is oblivious to the side-effects. In the end, the overall impact of image-processing operations on the perceived image quality or fidelity to the real world appearance is not thoroughly understood.

Evaluation of tone-mapping operators is an active research area [76, 77, 78, 79], which at the current stage is more focused on choosing correct psychophysical techniques than on providing clear guidance as to how existing operators should be improved to produce consistently high-quality images. Many existing evaluation methods treat each tested operator as a "black box" transfer function and compare its performance with respect to images produced by other operators, without explaining the reasons underlying human judgments. While some evaluation methods go one step further and attempt to analyze the reproduction quality of overall brightness, global contrast, and details (in dark and light image regions) [78, 79], but again they are focused on comparing which operator is better for each of these tasks. Those studies do not provide any deeper analysis as to how pixels of an HDR image have been transformed and what the impact of such a transformation is on desired tone-mapped image characteristics [80]. Another important question is how the outcome of the transformation depends on the particular HDR image content.

In a vast majority of perceptual experiments with tone mapping only one set of parameters per operator and per HDR image is considered in order to reduce the number of images that must be compared by subjects. The choice of the parameters may strongly affect the appearance of tone-mapped images and thus the operator performance in the experiment [81]. Another common problem is averaging the experimental results across subjects based on low-cross subject variability. This lack of variability can often be caused by the choices imposed on the subjects by the experiment design, which does not offer any possibility of adjusting the image appearance to subject's real preferences. The net result of published studies is that they often present contradictory results even if the same HDR images are used. Clearly, this suggests that the tone-mapping evaluation methodology should be improved.

Instead of the "black box" tone-mapping evaluation, there are some recent attempts of "bottom-up" approach in which the goal is to identify the low-level tone-mapping characteristics that lead to perceptually attractive images [81, 82]. For this purpose, the subjective preference and fidelity with respect to the real-world images is measured on an HDR display for images produced by a generic operator, whose characteristic and parameters are well understood. The goal of such research is to find some universal rules that facilitate a design of the operator that consistently produces preferred image appearance. For example, Seetzen et al. [82] found that for a given display peak luminance, there is a preferred level of contrast, which when exceeded leads to less preferred image appearance. The level of such optimal contrast increases with the display peak luminance. However, the preferred peak luminance should be below 6000–7000 cd/m^2, regardless of contrast, due to discomfort glare in dim ambient environments (the average surrounding luminance of 400–1200 cd/m^2 have been considered).

The correlation between image brightness and preferred contrast level has also been confirmed by Yoshida et al. [81], which also suggests that the use of these parameters to control tone mapping may be difficult for the user. Based on this observation, Yoshida et al. propose a better parameterization of a linear operator in logarithmic domain, in which parameters are more intuitive and can be partly estimated from image characteristics. Their operator is controlled by two parameters: *anchor white* and *contrast*. The *anchor white* parameter is approximately consistent across subjects and depends on images—it is set to a lower value if an image contains large self-luminous objects. The *contrast* parameter is more subjective, and therefore users should be allowed to adjust it. Yoshida et al. have shown that the parameters can be automatically estimated for their operator based on an image characteristic to obtain a "best guess" result. The *contrast* parameter can be predicted from the dynamic range of an image (images of higher dynamic range must be reproduced with lower contrast), and the *anchor white* parameter is related to the image key value (although the prediction can be unreliable if an image contains large self-luminous objects). The drawback of this approach is that the studied operator is very simple and does not deliver the image quality obtained using the state-of-the-art algorithms discussed in Section 6.1. Therefore, it remains to be seen whether the advanced operators can benefit from the proposed selection of parameters and an automatic estimation of their values as postulated in [81]. The problems of anchor white selection and overall image brightness control in terms of user preferences have been further addressed in [73].

Yoshida et al. have also investigated how the dynamic range and brightness of a display affects the preference for tone reproduction. For 14 simulated monitors of varying brightness and dynamic range, they did not find any major difference in the strategy the subjects use to adjust images for LDR and HDR displays. However, they noticed that the resulting images depend on a given task. If the goal is to find the best-looking image (preference), subjects tend to strongly enhance contrast (up to four times that of the original image contrast), even at the cost of clipping a large portion of the darkest pixels. On the other hand, when the task is to achieve the best fidelity with respect to a real-world scene, the subjects avoid clipping both in the dark and light parts of an image and they do not extend contrast much above the contrast of an original image. In both tasks, there is a tendency toward brighter images, which are achieved by over-saturating the brightest pixels belonging to self-luminous objects. Yoshida et al. have also compared the user's preference for displays of varying luminance ranges. The subjects prefer in the first order the displays that are bright, and in the second order, the displays that have low minimum luminance. Again, while such findings give useful insights how basic image display parameters affect the perceived image fidelity and preference, their integration to advanced tone-mapping operators is still an open research question.

6.3   OBJECTIVE EVALUATION OF TONE MAPPING

In this section, instead of subjective analysis, an objective perceptual metric is considered to help in understanding how particular image characteristics, such as contrast or brightness, are distorted by tone mapping with respect to the original HDR image. While objective metrics are usually less precise than data derived directly in psychophysical experiments, their big advantage is that a huge volume of images can be efficiently analyzed. This is particularly important in tone mapping where image characteristics affect the tone-mapped image appearance even if the same operator is used with consistently selected parameters [81].

The metric presented in this section is concerned with one well-defined suprathreshold distortion: contrast compression due to tone mapping, and uses the knowledge of human visual system to determine the perceived amount of such compression and to estimate the impact of such distortions on perceived image quality. In the following section, contrast distortions due to tone mapping are characterized and then the analysis of such distortions is presented for selected tone-mapping operators discussed in Section 6.1.

6.3.1   Contrast Distortion in Tone Mapping

All successful tone-mapping operators balance the tradeoff between plausible reproduction of the luminance range and preservation of details. One can argue that the *photographic tone reproduction* operator [65] best reproduces global contrast, while the *gradient domain compression* [74] operator best preserves details. However, the accuracy of such statements may depend on the particular HDR image, and as concluded by evaluations of tone-mapping operators [79,78], it is difficult for a tone-mapping operator to be well suited to all types of images. Regardless of technique, each tone-mapping operator introduces a degree of distortion into the resulting LDR tone-mapped image. Drawing conclusions from previous evaluations and general observations, two major contrast distortions can be identified that result from tone mapping:

Global contrast change. The ratio between lightest and darkest areas of the HDR is reduced in the LDR,

Detail visibility change. (Textures and contours) the high-frequency contrast of the HDR image becomes less prominent, disappears, or becomes exaggerated in the LDR.

A significant global contrast change is undesirable not only for esthetic reasons, but also because of changes in image understandability, despite good detail visibility. Certain

FIGURE 6.9: *Physical contrast* information in an LDR image (left) includes both visible and invisible details (middle). To discern only *perceived contrast* in the real-world conditions the detail visibility scaling into the Just Noticeable Differences (JND) units is required (right).

specialized tone-mapping operators assign a wider dynamic range to detail regions to preserve textures and contours, which results in a narrower dynamic range available for global luminance changes, decreasing the ratio between lightest and darkest areas. Detail visibility change occurs either because a region becomes entirely saturated or because an area is mapped to very few or very low brightness levels. The second case is especially interesting from the perceptual point of view, because the physical contrasts still exist in the LDR image, however the details are invisible to the human observer (refer to Fig. 6.9).

The goal of the objective metric is to determine the apparent distortion in detail visibility and global contrast change, that were introduced during the tone mapping of the HDR image, with the focus on the luminance compression aspect of the operators. Instead of analyzing particular algorithms one by one, the tone mapping is considered as an unknown transformation applied to the luminance of an HDR image, resulting in an LDR image. The output of the metric consists of a single value representing the global contrast change factor and a map representing the magnitude of change in detail visibility. The units of the detail visibility map are JND, which allows us to consider the visibility in the areas of an image and also permits us to use this information for potential perceptually-based corrections [83, 84].

For the details on the metric design, the readers are referred to [83].

6.3.2   Analysis of Tone-Mapping Algorithms

In this section, the aforementioned objective metric is used to analyze the performance of eight tone-mapping methods in terms of global contrast change and detail visibility

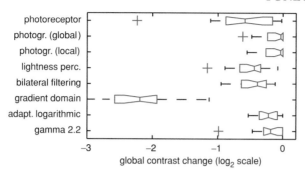

FIGURE 6.10: The influence of various tone-mapping operators on the change of the global contrast. The negative values denote the decrease in global contrast and 0 means no change. The red bars show the median, whiskers denote 25th and 75th percentile of data, and the red crosses are outliers.

change. The analysis was performed on a set of 18 HDR images with an average dynamic range of approximately 4 orders of magnitude and a resolution between 0.5 and 4 mega-pixels. The set contained a variety of scenes under different lighting conditions and included panoramic images. The following tone-mapping algorithms have been tested: global (spatially uniform)—*gamma correction* ($\gamma = 2.2$), *adaptive logarithmic mapping* [64], *photographic tone reproduction (global)* [65], *photoreceptor* [66] (Section 6.1.1); and local (detailed preserving algorithms)—*photographic tone reproduction (local)* [65] (Section 6.1.3), *bilateral filtering* [68] (Section 6.1.3), *lightness perception* [14] (Section 6.1.4), *gradient domain compression* [74] (Section 6.1.5). The tone-mapped LDR images were obtained either from the authors of these methods or by using publicly available implementations *pfstmo* (refer to Chapter 10). Tone-mapping parameters were fine tuned whenever default values did not produce satisfactory images.

The results of the global contrast change analysis are summarized in Fig. 6.10. There is an apparent advantage of the *photographic tone reproduction (local & global)* methods in conveying the global contrast impression almost without any change. These methods were also among the top rated in other studies [78, 79]. In contrast, the *gradient domain compression* causes a severe decrease in the global contrast. Other local methods perform moderately. Particularly, in the case of the *lightness perception* model the decrease of global contrast is caused by the optimization of difference in luminance between the frameworks. The superior performance of the global methods is traded for less efficient reproduction of details as observed in the further analysis.

The detail visibility change has been analyzed for two cases: the loss of detail visibility and the change in the magnitude of the detail visibility. The loss of detail visibility describes the situation in which details have been visible in the HDR image but are not perceivable in

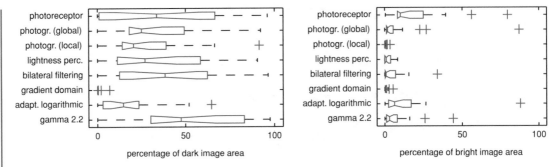

FIGURE 6.11: The influence of various tone-mapping operators on the loss of the detail visibility. The analysis is split into dark (left) and light (right) image areas. The percentage denotes the part of the dark/light image area in which details have been visible in the HDR image but are not perceivable in the tone-mapped image.

the tone-mapped image. The change in the magnitude of the detail visibility is considered only in the areas in which the details are visible both in the HDR and in the tone-mapped image. The average decrease and increase of the visibility are calculated separately. Following previous studies [79], the analysis has been further split into the dark and light image areas. To segment these areas, 33% of the darkest pixels in an image has been assigned to the dark area, and 33% of the brightest pixels to the light area. The results are summarized in Figs. 6.11 and 6.12. The results of the increase in detail visibility are not shown because they can be only observed for the *gradient domain compression.*

The analysis of Fig. 6.12 indicates that the dynamic range compression and the change in luminance levels lead to a decreased perception of details in the case of all operators. The magnitude of change, however, is in most cases below 1 JND. This means that the loss of detail visibility, largely observed in Fig. 6.11, is unlikely caused by the stark luminance range compression, but rather even a minor compression causes the magnitudes of details to drop below the visibility threshold. This would suggest that a minimal correction is sufficient to restore the visibility. The detail preserving tools implemented in local tone-mapping methods seem to perform well in light image areas; however, the dark image areas are often not well reproduced with the exception of the gradient domain compression and the adaptive logarithmic mapping. Notably, the *adaptive logarithmic mapping*, which is a global operator, preserves details exceptionally well in dark image areas. This advantage comes at the cost of a slightly higher loss of details in light areas. The *lightness perception* tone mapping performs on par with other local methods, being slightly advantageous in light image areas. The *gradient domain compression* is particularly interesting because the results of this detail preserving method indicate both the

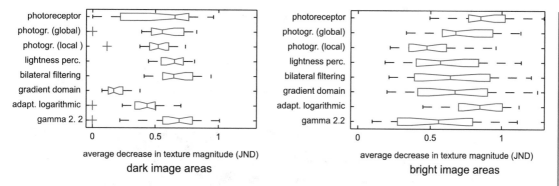

FIGURE 6.12: The average decrease of the magnitude of detail visibility caused by the analyzed tone-mapping operators. The analysis is split into dark (left) and light (right) image areas. The average is calculated over the parts where details are visible both in the HDR and in the tone-mapped image. 0 denotes no change in visibility and 1 JND denotes a visible change.

increase and decrease in detail visibility while at the same time the visibility of any details is not lost. Such behavior indicates good performance of the contrast transfer function which attenuates large contrasts and increases the small ones as explained in Section 6.1.5.

Overall, the better performance of the global tone-mapping operators in the analysis of global contrast change is not surprising. However, the performance of the algorithms in terms of detail visibility change is very unstable across the test images, and there is no obvious winner of the evaluation. Interestingly, the enhancements required to improve the results do not necessarily need to be strong. While the discovery of a new universal operator seems unlikely, such analysis motivates the development of enhancement algorithms that could restore the missing information in tone-mapped images based on their HDR originals. Such enhancements can be obtained using colors [83] or carefully shaped countershading profiles [84].

6.4 TEMPORAL ASPECTS OF TONE REPRODUCTION

The tone-mapping algorithms discussed so far have been designed for static images, what in principle means that the illumination conditions and luminance levels are assumed constant. In the HDR video, as also in the natural world, the illumination changes. The human eyes adapt their response range to the current ambient light level. Normally, the adaptation processes are mostly not noted because the changes in the illumination during the course of day and night are very slow. Sudden changes, however, cause visible loss in the sensitivity as illustrated in Fig. 6.13. For instance, when on a sunny day one immediately enters a dark theater, the interior

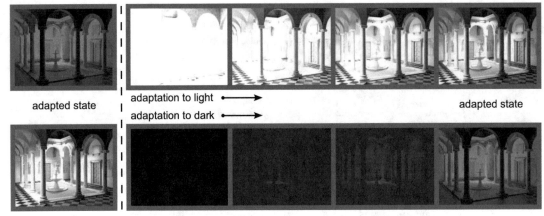

sudden change in illumination

FIGURE 6.13: Visual experience in certain time intervals during the temporal adaptation to light and to dark caused by a sudden change in illumination. The visibility improves with time because the response range of photoreceptors adjusts to the medium illumination in the scene.

is at first dark and no details can be discerned—only after several seconds the silhouettes of objects start to appear.

The adaptation of human eyes to light is a temporal process. The precise time course of adaptation can be measured and is shown in Fig. 6.14. The plots start with a sudden change in illumination which results in loss of sensitivity. The sensitivity of both rods and cones

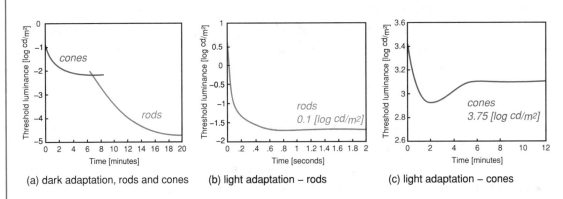

(a) dark adaptation, rods and cones (b) light adaptation – rods (c) light adaptation – cones

FIGURE 6.14: Time course of dark adaptation (a) and light adaptation (b,c) as a function of sensitivity. Higher threshold values indicate that the eyes are not well adapted; thus, the sensitivity is low. Dark adaptation was to complete darkness, light adaptation to the specified luminance levels. Redrawn from [85].

progresses asymptotically. During the dark adaptation, the process of cones is faster but cones soon reach their maximum sensitivity. The sensitivity level is for a moment constant because the rods still have not recovered from the strong illumination. With time, rods dominate the vision and continue the adaptation process until the maximum sensitivity. The light adaptation in the scotopic range is extremely rapid and nearly 75% of the process is accomplished in first 400 ms. The cone system adapts to light much slower and requires about 3 min to reach the maximum sensitivity which then slightly decreases. Due to their asymptotic nature, the adaptation processes are often approximated with the exponential function.

Similarly as in the natural world, the luminance values in the HDR video can significantly change from frame to frame and cause unnatural brightness changes in the tone-mapping results. To prevent this, tone-mapping operators for video implement mechanisms that are similar to the adaptation processes in human eyes. The goal of these mechanisms is twofold: in principle they guarantee natural appearance of light changes in the video stream, but also they assure the temporal coherence between frames. The temporal coherence is an important issue because small changes in the luminance distribution between video frames often influence the brightness of tone-mapping result what in turn causes undesired brightness oscillations in the displayed HDR video stream. While the first goal may require faithful modeling of temporal adaptation processes in human vision, the temporal coherence can be achieved even by simple models [86].

In the luminance-based tone-mapping algorithms, the light adaptation is usually modeled using the adapting luminance term given in Eq. (6.2). To achieve temporal coherence for video, instead of using the actual adapting luminance Y_A for the displayed frame, a filtered value $\bar{Y}_A$ is used. In most implementations, the value of $\bar{Y}_A$ changes approximately according to the adaptation processes in human vision, eventually reaching the actual value if the adapting luminance is stable for some time. The adapting luminance is filtered using an exponential decay function [87]:

$$\bar{Y}_A^{\text{new}} = \bar{Y}_A + (Y_A - \bar{Y}_A) \cdot (1 - e^{-\frac{T}{\tau}}), \tag{6.8}$$

where T is the discrete time step between the display of two frames, and τ is the time constant describing the speed of the adaptation process. Depending on the required faithfulness to the actual adaptation processes, the time constant can be one for all light conditions, or can be different for rods and for cones, or even may depend on the pre-adaptation processes [88]. Commonly chosen values for adaptation of rods and cones are as follows:

$$\tau_{\text{rods}} = 0.4 \text{ s} \qquad \tau_{\text{cones}} = 0.1 \text{ s}, \tag{6.9}$$

and if only the temporal coherence is required, the τ_{cones} constant is used. Furthermore, the time required to reach the fully adapted state also depends whether the observer is adapting to light or dark conditions. The values in Eq. (6.9) describe the adaptation to light. For practical reasons, the adaptation to dark is not simulated because the full process takes up to tens of minutes as shown in Fig. 6.14. Instead, the adaptation is most often performed symmetrically neglecting the case of a longer adaptation to dark conditions. The complete tone mapping solution for HDR video can be found in [41] and in [27].

6.5   CONCLUSIONS

In view of the increasing availability of the HDR contents, the problem of their presentation on conventional display devices is highly recognized. Different goals and approaches led to the development of versatile algorithms. These algorithms have different properties which correspond to the specific requirements and applications. Furthermore, due to the temporal incoherence certain methods cannot be used for the tone mapping of video streams. A universal method has not been found so far; therefore, the choice of the tone-mapping method should be based on the application requirements. It is also not clear how to evaluate tone-mapping operators in terms of image quality because their performance depends strongly on the choice of parameter values and the actual HDR image content. The development of robust methods that could be used for the automatic parameter tuning to obtain desirable image appearance is still an open research question. Also, the problem of color appearance, which depends a great deal on luminance level, has not been researched too deeply.

With respect to the HDR video streams, the choice of an appropriate tone-mapping method is usually a tradeoff between the computational intensity and the quality of dynamic range compression. The quality here is mainly assessed by a good local detail visibility. The global tone-mapping methods are very fast, but often leads to the loss of local details due to an intensive dynamic range compression. Such methods should be used whenever high efficiency is the main requirement of the target application. The adaptation mechanisms can be used to select the range of luminance values which should obtain the best mapping. However when the quality is insufficient, local tone-mapping methods are necessary. The local detail enhancement methods provide a good improvement to the global tone-mapping methods still achieving good computational performance.

The photometrically calibrated HDR video streams allow for the prediction of the perceptual effects such as reduced visual acuity and lack of color vision for the rod vision, motion blur, and glare (see Fig. 9.2 and refer to Section 9.1). Such effects are typical to everyday perception of real-world scenes, but do not appear when observing a display showing a tone-mapped HDR video. Prediction of such effects and their simulation can increase the realism of the presentation of HDR contents. On the other hand, such a prediction may also be used to identify situations when a real-world observation of scene would be impaired and to hint the tone-mapping algorithm to focus on the good detail reproduction there.

In Chapter 10, we provide more information on the *pfstmo* software package [33] containing implementations of many state-of-the-art tone mapping described in this chapter. The package is available under the URL:

`http://www.mpi-inf.mpg.de/resources/tmo/`

CHAPTER 7

HDR Display Devices

In recent years, we witness important developments in HDR display and projection technology. In this chapter, we discuss basic requirements imposed on this technology from the standpoint of selected characteristics of the human visual system (HVS), which are important in image perception. We also give examples of selected technical solutions used in HDR display technology and we discuss their merits and limitations.

7.1   HDR DISPLAY REQUIREMENTS

An ideal display device should not introduce any visible image quality degradation with respect to the observation conditions for the real-world scenes. This means that technical capabilities of such an ultimate display device should outperform the limitations imposed by the HVS. The following characteristics of the HVS are important in image perception.

- The contrast sensitivity function (CSF), which determines the HVS ability to resolve image patterns of various spatial frequencies. The display resolution should enable us to reproduce all spatial frequencies that can be seen by the human eye. The CSF for luminance and chrominance patterns should be considered, but in practice the former one is the limiting factor because of higher the HVS sensitivity to luminance.

- The threshold-versus-intensity (tvi) function, which describes the just noticeable difference (JND) of luminance and chrominance that can be detected in the image for given luminance adaptation conditions. In fact the tvi-function can be derived by extracting the maximum sensitivity values from the family of CSFs, which are measured for various background luminances. The quantization step in luminance and chrominance encoding in the display should be below one JND to avoid contouring (banding) artifacts that are visible in particular when reproducing smoothly changing image patterns.

- The luminance range that can be simultaneously seen by the HVS for given adaptation conditions. The display dynamic range determined by the minimum and maximum luminance values should match the HVS capabilities. The dynamic range decides about the maximum global contrast that can be reproduced by the display.

- *Color gamut seen by the HVS.* The display primaries determine the actual gamut that can be reproduced in displayed images. The gamut also changes with the display dynamic range.

- The field of view which affects the immersion experience and decides upon adaptation conditions. The visual field measured for binocular human vision extends over 200° (width) × 135° (height).

An important question arises what are the limitations of current display technology in terms of matching the just listed HVS characteristics that are important in image perception?

The best match can be observed between the CSF and display resolution. Image patterns of spatial frequency up to 50 cycles-per-degree (cpd) can be still reproduced on the high-definition (HD) displays featuring the image resolution of 1920×1080 pixels for the observer distance larger than 5 screen heights. Since even the high-contrast luminance patterns of this spatial frequency are barely visible by the human eye, it can be considered that the HD display technology matches the HVS capabilities in terms of spatial pattern reproduction. In practical TV viewing conditions with significant ambient lighting, it is often assumed that only patterns up to 30 cycles-per-degree (cpd) can be seen and thus 3 screen heights is the recommended watching distance to take the full advantage the HD image resolution. Note that the watching distance effectively defines the field of view covered by the display. The HD resolution is also sufficient for brighter displays that might be available in the future because the shape of CSF does not change significantly for adaptation luminance above 1000 cd/m^2 (refer to Fig. 4.5).

The quantization step in encoding physical luminance and chrominance values, which can be reproduced by the display, obviously depends on its dynamic range. As we discussed in Section 2.3 the HVS can simultaneously see the luminance range up to 4–5 orders of magnitude. For natural scenes, which feature even wider dynamic range, an appropriate subset is selected through complex adaptation mechanisms. Under display observation conditions, such adaptation strongly depends on ambient light in the surrounding environment as well as light emitted by the display itself. The resulting adaptation anchors the range of simultaneously visible luminance and determines the minimum and maximum luminance values that can be seen. Seetzen et al. [82] have found for a darkened room the maximum luminance values that can be comfortably seen is of the order 6000–7000 cd/m^2. Under such conditions, the minimum luminance that can be seen is of the order of 0.01 cd/m^2. In practice, the display black level is affected by the ambient light reflected in the display screen. As predicted by the JND-space encoding (refer to Section 5.1.6) for such bright displays the quantization artifacts are easier to see, which means that 8-bit encoding of such wide luminance range is not sufficient, and at least 10-bit encoding may be required (in fact it is safer to assume even smaller quantization error as offered by 12–16-bit encoding). More than 8 bits is also required for chrominance encoding, in particular, for blue and purple colors for the highest luminance levels.

For such a display specification and assumed dark environment the HVS performance will be close to optimal and further increase of the display luminance range and as well as reduction of the quantization error cannot improve this performance. In practice, modern displays rarely meet such requirements: 8-bit quantization is predominant and the ANSI contrast numbers as measured for black and white checkerboard are of the order from 1:50 to 1:500, which is far from desirable 4–5 orders of magnitude. The contrast specification provided by many display manufacturers is based on luminance measurements for the full-on and full-off screens, which leads to strongly exaggerated contrast values because light leakage from neighboring bright to dark regions is not accounted for.

Recently, the so-called HDR display devices have been developed whose specification approaches limits imposed by the HVS in terms of reproduced contrast and quantization error. Two basic technologies have been used to achieve this goal: dual modulation and laser projection. Dual modulation relies on optical multiplication of two independently modulated representations of the same image. Effectively, the resulting image contrast is a product of contrast achieved for each component image, while only standard 8-bit drivers are used to control pixel values. In laser projectors, the laser light is scanned over the screen surfaces with light intensity directly modulated using 12–16-bit drivers. In the following sections, we briefly describe both technologies.

7.2   DUAL-MODULATION DISPLAYS

In the basic design of a dual-modulation display, the input HDR image is decomposed into low-resolution *backlight* image and high-resolution *compensation* image as shown in Figure 7.1. The requirement of precise alignment of pixels between the two images can be relaxed due to blur in the backlight image, which does not contain high spatial frequencies. Therefore, as the result of optical multiplication between backlight and compensation images, the achieved global contrast (low spatial frequency) is a product of contrasts in both images, while the local pixel-to-pixel contrast (high spatial frequency) arises only from the compensation image. While this is not a problem for low contrast image patterns, which are successfully reproduced even on traditional single-modulator LDR displays, local pixel-to-pixel contrast reproduction in the proximity of high-contrast edges may not be precise. Fortunately, the veiling glare effect caused by imperfections of the human eye optics leads to polluting retinal photoreceptors, which represent dark image regions with parasite light coming from bright regions. Thus, the veiling glare makes impossible to see sharply such local patterns of high contrast, which effectively

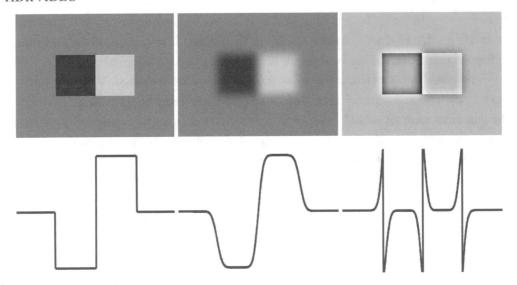

FIGURE 7.1: Decomposition of the source image (left) into the low-resolution backlight image (middle) and the high-resolution compensation image. Images courtesy of Gerwin Damberg, Helge Seetzen, Greg Ward of Dolby Canada and Wolfgang Heidrich and Lorne Whitehead of the University of British Columbia. Reproduced from [89] with permission by The Society for Information Display.

means that they do not have to be reproduced by the display. Obviously, high contrast between more distant image regions, which can readily be seen be the eye, is faithfully reproduced.

The backlight and compensation images require special image processing so that their multiplication results in the reconstruction of the original HDR image. The goal of such image processing is to account for different image resolutions and the optical blur in the backlight image. For this purpose, the point-spread function (PSF) characterizing this blur should be modeled for all pixels of the backlight image. The overall flow of image processing in the dual-modulation display architecture is shown in Fig. 7.2. At first the square-root function is used to compress the luminance contrast in the input HDR image and then the resulting luminance image is downsampled to obtain the low-resolution backlight image. In the following step, the PSF is modeled for every pixel of the backlight image, which simulates the light field (LFS) that effectively illuminates the high-resolution modulator. By dividing the input HDR image by the LFS the high-resolution compensation image is computed. Since the compensation image is 8-bit encoded, some of its regions may be saturated, which results in undesirable detail loss. Such saturation errors are analyzed and a close-loop control system is used to locally increase the intensity of pixels in the backlight image to prevent such saturation. Figure 7.1 shows an example of backlight and compensation images resulting from such image processing.

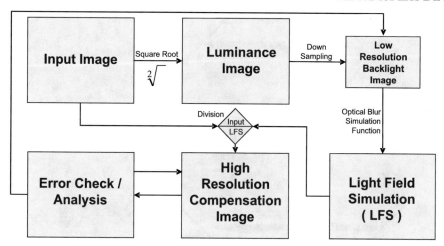

FIGURE 7.2: Image processing flow required to drive low- and high-resolution modulators in HDR projection/display system. Image courtesy of Gerwin Damberg, Helge Seetzen, Greg Ward of Dolby Canada and Wolfgang Heidrich and Lorne Whitehead of the University of British Columbia. Reproduced from [89] with permission by The Society for Information Display.

The dual-modulation technology has been successfully used to build HDR projection [89] and display systems [2], [27, Chapter 14]. In both cases standard 8-bit LCD panels have been used for modulation of the compensation image, and major construction differences come from realization of the backlight modulator. For the projection system developed by Damberg et al. [89] a passive low-resolution LCD modulators with a fixed light source has been used. Figure 7.3 illustrates extensions introduced to a standard projection system with three transmissive LCD panels modulating RGB channels. Three low-resolution transmissive LCD panels have been placed next to the existing high-resolution panels. Such a design enables very faithful color reproduction and the amount of blur can be controlled by changing the distances between each pair of low- and high-resolution RGB panels. The low resolution of the backlight modulator leads also to a better efficiency of light transmission because density of electronic components and other blocking elements can be reduced [89]. Damberg et al. reported that in their projection system they achieved 2,695:1 contrast, which is only by 5% lower than the theoretical product of contrast reproduced by the low (1:18) and high (1:155)resolution modulators. The authors experimented also with other projector architectures by changing the order of high and low-resolution panels, or using just a single low-resolution luminance modulator, which is placed between the X-Prism and the lens system i.e., after the recombination of light modulated by the three high-resolution RGB channels. The generic HDR projector architecture as proposed in [89] can be also used for other projection technologies based on digital micro-mirror devices (DMD) and liquid crystal on silicon (LCoS).

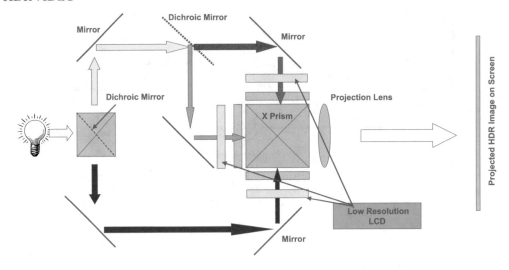

FIGURE 7.3: Example of implementation of a three-LCD projector augmented with three low-resolution backlight modulators for RGB color channels. Image courtesy of Gerwin Damberg, Helge Seetzen, Greg Ward of Dolby Canada and Wolfgang Heidrich and Lorne Whitehead of the University of British Columbia. Reproduced from [89] with permission by The Society for Information Display.

For HDR displays passive modulators have been used as well, but much better energy efficiency has been achieved using active backlight modulators based on a matrix of independently modulated light emitting diodes (IMLED) [2]. Interestingly, such spatially-varying backlit device is 3–5 times power efficient than uniform light employed in conventional LCD displays of similar brightness [27, Chapter 14]. Also, the color gamut can be significantly expanded if different color LED (e.g., integrated RGB LED packages) are used instead of white light commonly used in conventional LCD displays. Brightside Technologies developed a number of prototype HDR displays, and their recent DR37-P model features the maximum luminance up to 3000 cd/m^2 and almost perfect black level of 0.015 cd/m^2, which is limited only by parasite lighting that may leak from neighboring active LEDs. This gives remarkable 1:200 000 global contrast while the measured ANSI contrast for the black and white checkerboard pattern reaches 1:25 000. BrightSide DR37-P is the full HD 1920 × 1080 display with 37″ screen diagonal. For the backlight device 1200 LEDs have been used, which form a symmetric hexagonal grid.

The use of IMLED matrices as backlight devices becomes more and more popular in modern LCD TV sets. Just recently LG Philips introduced on the market a novel Local Area Luminance Control in their 47″ TV sets with LED backlight. Also, Samsung developed Local Dimming LED technology. For these technologies, the cooling problem is the main issue that prevents installing more powerful LEDs in these displays and making them full-fledged

HDR displays. However, given that the number of lumens per watt [lm/W] in modern LEDs increases at a higher rate than the Moore's law, upgrading Philips and Samsung technology to the specification (in terms of contrast and luminance range) similar to the BrightSide's HDR display may be a matter of relatively short time. Also, it can be envisioned that with progressing miniaturization of small high power light source arrays the active backlight technology will also be employed for future projection systems.

Overall the dual-modulation technology offers an inexpensive way of doubling the bit-depth controlling the luminance or color channels, and achieving remarkable global contrast and the maximum luminance specifications for HDR projection and display systems.

7.3   LASER PROJECTION SYSTEMS

Laser projection technology is a promising alternative for displaying HDR images. For example, the Scanning Laser Display Technology developed by JENOPTIK GmbH [90] (refer to Fig. 7.4) employs 12–16 bit image encoding and directly reproduces bright and dark pixels through modulating the amplitude of RGB laser beams. Acousto-optical modulators are used to transform the RGB video signal into optical information. Then the three modulated laser beams are combined into one collinear beam, which is transferred to the projection head (scanner unit) using an optical fiber, whose length can be up to 30 m. This separation of large laser system from the compact scanner unit is very convenient for many applications. The modulated light arriving to the scanner unit is deflected in the horizontal direction using a rotating head with

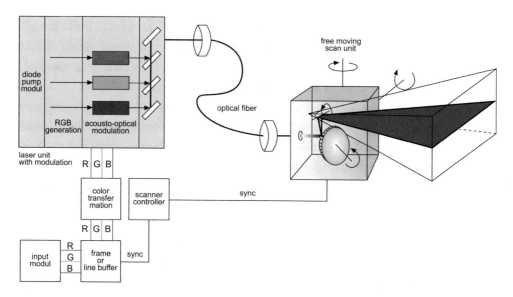

FIGURE 7.4: Scanning laser projection display developed by JENOPTIK GmbH. Redrawn from [90].

25 mirror facets, which results in the scan angle of about 26°. The vertical deflection of image scanlines is performed using a galvanometer mirror, which allows a full deflection angle of about 20°. The flying spot of the laser beam results in the very smooth transition between neighboring pixels (absence of visible pixel boundaries). The image resolution can easily be enhanced, motion blur is practically invisible due to fast line scan time, and native bit depth of amplitude modulation is very high. The image can be easily projected on curved surfaces because of large depth of sharpness ranging from 5 to 50 m and good color convergence. The full on/full off contrast ratio is higher than 1:100 000, which in simulation applications enables day and night simulation with the same equipment. Another advantage of laser projection technology is enlarged color gamut due to more saturated primaries determined by the wavelengths of lasers. With extended contrast offered by the projector this leads to more saturated and vivid colors. The fixed laser wavelengths and power control enables good temporal stability in color reproduction.

The main disadvantage of laser projection technology is moderate peak luminance level, which is limited by the power of laser diodes. Another limiting factor is the high cost of major system components such as lasers and light modulators. There is some hope that the cost barriers will be overcome with increasing interests in laser television (TV). In recent years, the rapid progress in the development of 1-W and higher power RGB lasers can be observed. Also, after a successful application of digital micromirror device (DMD) technology in projection systems, new generation of microelectro-mechanical systems (MEMS) have been successfully tested as linear light modulator arrays. Grating light valve (GLV) and grating electro-mechanical system (GEMS) technologies are much cheaper in manufacturing than the DMD devices and much faster (× 1000) in switching between their states. Effectively this enables to build just a high-resolution column of pixels which through laser scanning and deflection of the reflected beam can reconstruct an image of very high resolution. For example, the GLV switching speed of 20 ns is sufficient to build even four such images during a conventional video frame, which enables to improve the bit-depth for color channels using temporal dithering approach (effectively a smaller quantization step can be achieved through averaging of subsequently displayed images by the HVS). For example, GEMS-based laser projection system demonstrated by the Eastman Kodak Company featured superb image quality with wide color gamut, reduced motion artifacts, HD resolution, and high native bit depth [91].

7.4   CONCLUSIONS

In this chapter, we have outlined recent developments in HDR display technology. In coming years, a rapid development of such technology can be anticipated and virtually every month brings some announcements from the industry on launching on the market new HDR projection and display devices. Digital cinema applications are the driving force for the professional market of HDR projectors. For the consumer market, the dual-modulation technology with LCD displays becomes particularly attractive with dropping prices of high-power LEDs and improving their luminous efficiency. Also, integrated circuits (IC), which are capable of steering larger and larger LED matrices, are actively developed due to increasing demands from other industries e.g., automotive. It seems that at the current stage every major manufacturer is preparing for launching LCD displays based on some form of local dimming technology to deepen the black level of the display. The availability of energy efficient LEDs, which feature high luminous power, will improve the image reproduction in bright regions without imposing excessive demands on the display cooling system. In such situation, the main problem, which we discuss in the next chapter, is to deliver HDR content that fully can exploit the capabilities of modern display technology.

CHAPTER 8

LDR2HDR: Recovering Dynamic Range in Legacy Content

Historically CRT display devices have been predominantly used to render digital content, and their capabilities in terms of reproduced contrast (typically up to 1:100) and luminance range (typically 1–100 cd/m^2) have a profound impact on image and video formats, which have been specifically tailored for these capabilities. In such display-referred LDR formats information for every pixel is encoded directly in a ready-to-use format with the goal that reproduced images should "look good" on any device and should not require any further processing. This strategy of digital content storage turned out to be far from optimal with increasing diversity of display and projection technologies, which are capable of reproducing wider contrast ranges (typically up to 1:400 for modern LCD and plasma displays), feature more profound black levels and maximum luminance values, and improve image sharpness. For these technologies, the precision deficiency in the existing image and video formats may result in visually disturbing quantization artifacts, which modern LCD displays can practically eliminate through on-line decontouring and bit-depth expansion (refer to Section 8.1).

Such simple means are not sufficient any more for full-fledged HDR displays such as Brightside DR37-P (refer to Section 7.2). For such displays recovering HDR information in legacy LDR images and video is required, which is often called inverse tone mapping or simply LDR2HDR. The main problem here is to find nonlinearity of contrast compressing function applied to each LDR image and to overcome the quantization errors in the recovered HDR image. Another important problem is restoring (inpainting) image details in highlights, light sources, and deep shadows, which are typically clipped in LDR images, but can easily be displayable on HDR displays.

The LDR2HDR problem can formally be stated as the reproduction of real-world luminance values for every pixel in an LDR image. Such stated problem without making extra assumptions concerning the image capturing system as well as captured scene itself is ill-posed and in general case cannot be solved in an automatic way. The first unknown factor on the way of light from the scene toward the camera sensor is the lens system. The direct illumination in the scene that should be registered for each sensor pixel is polluted through indirect lighting scattered in the camera optics due to veiling glare and lens flare effects. Another important factor is the camera sensor response, which can be a complex-shaped function that is difficult to recover faithfully from a single LDR image. The captured image is polluted by the sensor noise, which makes dark pixels less reliable. Information is lost completely for excessively exposed and thus saturated pixels. Finally, the raw sensor image usually undergoes sophisticated image enhancing, sharpening, and tone mapping (possible further pixel clipping both in dark and light image regions) using proprietary and generally unknown algorithms before its encoding in any standard format. All these factors make the task of precise scene luminance map reconstruction very difficult. In practice, the goal of LDR2HDR processing is formulated less strictly in terms of achieving visually plausible image appearance on an HDR display. We summarize existing solutions, which can contribute to dynamic range expansion and are suitable for legacy video and images.

- Bit-depth expansion and decontouring techniques (Section 8.1),
- Reversing tone-mapped curve in LDR images (Section 8.2),
- Recovering camera response curve from a single LDR image (Section 8.3),
- Recovering (inpainting) image details in saturated shadow, highlight, and light source regions (Section 8.4),
- Handling video on-the-fly (Section 8.5),
- Taking advantage of image artifacts due to acquisition problems for recovering useful HDR information (Section 8.6).

In the following sections, we discuss the problem of upgrading the existing LDR image and video content to make it suitable for HDR display and projection systems such as those discussed Chapter 7. We focus mostly on restoring luminance component. We do not cover another important problem of extending color gamut, e.g., extending chromaticity values toward higher saturation, without changing the hue as required for projectors and displays with color primaries based on lasers and LEDs. Such problems are partially discussed in the literature on gamut expansion. Also, we do not address the problem of quantized colors restoration, which is in particular a difficult task when the quantization method is not known a priori [92].

8.1   BIT-DEPTH EXPANSION AND DECONTOURING TECHNIQUES

In many traditional LDR imaging pipelines, usually based on 24 bits/pixel, there are often some components which impose limitations on the number of bits per pixel. For example, in DVD applications tailored for the CRT displays the compressed image quality is effectively equivalent to 6-bit signal because information from the two least significant bits in the original 8-bit encoding is usually removed due to the quantization errors. Note that for new generation LCDs, which are very bright and feature little noise, 10-bit accuracy of internal processing is often required so that the analog signal, which steers the liquid crystals can reproduce the smallest contrast details that the human eye can perceive. It should be noted that this conservative requirement concerns only the spatial frequencies of 4–8 cpd and for other spatial frequencies a lower number bits is sufficient (e.g., 4 bits for spatial frequencies greater than 27 cpd) [93]. This is an important change with respect to the CRT technology, which required only 6–8-bit accuracy due to lower luminance levels (lower eye sensitivity for contrast), more blurry and noisy pixels (more visual masking suppressing the visibility of low contrast details). Excessively limited bit-depth obviously results in loss of low amplitude details that are below the corresponding quantization error, but could be potentially visible on a high-quality display device. Another visual consequence of limited bit-depth is contouring, which forms false contours (also called banding artifacts) in smooth gradient regions (such contouring for chromatic channels is often called posterization). We discuss two types of techniques designed to reduce these artifacts.

- Pre-processing techniques in which the high-bit depth reference image is available and it can be used to modify the low bit-depth image version by adding noise or amplifying its low amplitude features, so that this information can survive the image quantization and can be recovered at the display stage. We describe briefly *bit-depth expansion (BDE)* and *compander* techniques, which belong to this category.

- Post-processing techniques for which the only available information is the low bit-depth image and the main goal is removing existing contouring artifacts (decontouring). We outline adaptive filtering, coring, and predictive cancellation techniques, which are examples of post-processing techniques. These techniques are often implemented in hardware installed in modern LCD and plasma TVs to achieve real-time performance.

BDE techniques are designed specifically to achieve higher perceived bit depth quality than it is physically available. As in dithering techniques, usually the BDE techniques rely on adding imperceptible spatiotemporal noise to an image prior to the quantization step. Intensity averaging in the optics of display and human eye leads to recovering information below the quantization step. Modern BDE techniques tune a micro-dither amplitude to obtain a low-spatial frequency flicker from mutually high-pass spatial and temporal noise and achieve 10-bit perceived quality on 8 bit-driver LCDs [94]. In designing power spectral density and amplitude characteristics of the noise, it is important to take into account the knowledge of human visual system, so that the noise remains invisible. Otherwise, perceptible noise would not only degrade the visual quality, but additionally could mask the low amplitude image details, which is just the opposite effect to the fundamental goal of BDE techniques. The noise visibility can be kept under control by setting the noise amplitude below thresholds predicted by spatio-temporal contrast sensitivity function (CSF). Also, the spectral density noise characteristics can be effectively pushed to higher, less perceptible spatial, and temporal frequencies. Another factor that should be considered in evaluation of noise visibility is increasing the eye sensitivity to contrast for modern display devices due to much brighter images with respect to CRT devices.

Li et al. [59] propose a wavelet multiband technique for the compression of an high bit-depth image into an low bit-depth image and then the expansion of its dynamic range back (the so-called compander). The information loss is reduced by amplifying (pre-distorting) low amplitudes and high frequencies at the compression stage, so that they survive the quantization step to the 8-bit LDR image. Since the bit-depth expansion is a fully symmetric inverted process, the amplified signals are suppressed back to their initial level in the companded high bit-depth image. The authors observe that their compander leads to a good quality reconstruction of HDR images based just on 8-bit LDR images, whose visual quality is also acceptable. However, it seems that this technique has more potential for HDR image compression rather than pure bit-depth expansion. We discuss the compression aspect of this technique in more detail in Section 5.4.2.

When higher bit-depth information is no longer available, which is often the case for legacy content, low-amplitude details cannot be reconstructed, and post-processing is focused on removing false contours [93, 95]. *Adaptive filtering* relies on smoothing contouring artifacts without introducing excessive blur to an image. For example bilateral filtering can be used for this purpose by removing from the image information of high frequency and low amplitude. This can be achieved by setting the intensity domain parameters of Gaussian filter tuned to expected contouring contrast and limiting the spatial Gaussian filter support to few neighboring pixels. *Coring* techniques are essentially based on the same principle, but offer more control over high-frequency details filtering through multiband image representation [96]. Filtering is applied only to a couple of high-frequency bands and its strength smoothly decreases towards lower frequency bands. In adaptive filtering and coring methods details of low amplitude and high frequency may be lost, which may affect the visual image quality. For example, excessive smoothing of the human skin texture may lead to its unnatural plastic appearance, which is highly undesirable effect for any commercial broadcasting and display system.

In *predictive cancelation*, the idea is to estimate the quantization error based on input low bit-depth image and compensating for this error prior to the image display. To achieve this goal, the low bit-depth image P undergoes low-pass filtering, which results in low-spatial frequency image L whose pixels have higher precision than in P due to averaging (refer to Fig. 8.1). Of course, this precision gain in L is obtained only for slowly changing signals, at

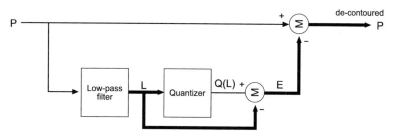

FIGURE 8.1: Predictive cancelation flowchart. Thick lines denote a higher-bit precision in the image representation. The de-contoured image P is submitted to a display device.

expense of original spatial resolution at P. Now, when the quantization operator Q with the same bit-depth accuracy as in P is applied to L, the difference $E = Q(L) - L$ approximates the quantization error inherent for P, but only for low spatial frequencies. Then, by subtracting the error E from P the most objectionable contouring due to slowly changing image gradients is removed. At the same time, potential contouring at higher spatial frequencies remains intact, but here the eye sensitivity to contrast is lower as predicted by the CSF. Also, in the high contrast image regions with significant high spatial frequency content (e.g., some texture patterns) visual masking can further help in hiding contouring artifacts.

Recently, Bhagavathy et al. [95] have proposed a multi-scale probabilistic dithering method, which comprises two main steps. At first, a multi-scale analysis on the neighborhood of each pixel determines the likelihood of banding for this pixel. A pixel is assumed to be a part of banding artifact, when the likelihood of banding is larger than a predefined threshold value at least one scale. Then banding reduction is performed for such a pixel by computing a local mean (floating-point) intensity in the pixel neighborhood, which is then probabilistically dithered and quantized as required for the new bit depth. The proposed method is less dependent on the proximity between adjacent false contours than methods relying on smoothing filters with predefined support such as predictive cancelation. On the other hand, the proposed method is sensitive on the preset threshold of banding likelihood, which is used to detect pixels contributing to banding artifacts.

All discussed BDE and decontouring techniques are optimized for much lower bit-depth expansion than required to accommodate HDR image and video content, so their adaptation to the LDR2HDR problem is an open research question. An exception is the compander technique, which has been successfully applied for dynamic range compression and expansion, but only in the context of static HDR images. Also, this method requires the HDR reference and strongly enhances low-contrast information in the compressed image, which may not be acceptable in some applications for which the fidelity of compressed image appearance is important. The decontouring techniques may have some potential for contrast boosting techniques described in the following section in particular for lower quality and low bit-depth legacy video and images.

8.2   REVERSING TONE-MAPPING CURVE

For high quality LDR images with a small amount of under- and over-exposed pixels, which do not contain visible quantization and compression artifacts, deriving inverse tone-mapping function, then transforming all pixel values using this function, and finally contrast expansion seems to be an easiest recipe to reconstruct the corresponding HDR images. A number of solutions presented in the literature adopted such a procedure [97, 98, 99, 100], and they differ mostly in the precision of inverse tone-mapping function derivation and the actual contrast expansion approach.

Akyüz et. al. [97] conducted a psychophysical study on an HDR display, in which they ranked the general preference for high quality HDR images and the corresponding LDR images with linearly/nonlinearly scaled contrast and brightness to fully exploit the dynamic range of an HDR display. Each source LDR image, which has been submitted for such scaling, has been selected as the best-exposed image from the pool of images merged through a multi-exposure technique into the corresponding high-quality HDR image. It turned out that the subjects similarly ranked the LDR images with linear contrast scaling as the corresponding HDR images. This may suggest that for a good quality LDR image simple contrast boosting may be sufficient for many scenes.

In another psychophysical study, Meylan et al. [98, 101] confirmed this observation for scenes featuring lower dynamic range. However, they argued that just a linear rescaling of images that are tone mapped for standard displays may lead to too bright images when displayed on an HDR display. They found that usually better results can be obtained by taking into account the actual image content and by diversifying contrast boosting for diffuse regions and highlights. In their inverse tone-mapping algorithm, they segment the diffuse and highlights image parts, which are then independently rendered with two different linear scaling functions r_1 and r_2. This way the lower part of display dynamic range is used to render the scene and the remaining part is allocated for visualization of highlights and light sources (refer to Fig. 8.2). The splitting point between these parts is decided based on the maximum diffuse white W_{in} in the input LDR image and assigned to this point display luminance value W_{out}. Parameter W_{out} should be adjusted on the basis of the image content to control the overall image brightness impression and can be a function of the size of highlight regions. In [98] Meylan et al. describe a psychophysical experiment in which they investigate the subject preference for various W_{out} choices. They found that for outdoor scenes the subjects preferred to allocate a rather small part of the dynamic range to specular highlights to achieve overall brighter image

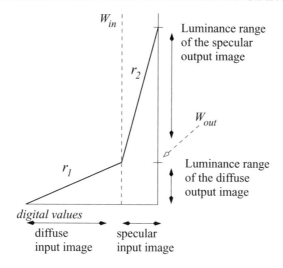

FIGURE 8.2: Display dynamic range allocation between diffuse and specular image parts. W_{in} refers to the maximum diffuse white in the input LDR image and W_{out} denotes the corresponding intensity in the output dynamic range enhanced image. Figure courtesy of Laurence Meylan of General Electric, Scott Daly of Sharp Laboratories of America, and Sabine Süsstrunk of Ecole Polytechnique Fédérale de Lausanne (EPFL). Reprinted from [102] with permission from SPIE.

appearance. For indoor scenes, better visual results were obtained when more dynamic range was allocated for highlights. Also, the percentage of specular pixels can be important (e.g., the sun reflecting in the water surface), in which case the subjects prefer dimmer images. In all tested cases boosting brightness of specular highlights led to more natural impression, which indicates that content-dependent inverse tone mapping may be favorable (refer to Fig. 8.3).

In the follow-up paper Meylan et al. [102] investigate an automatic algorithm for high-light detection and determination of the maximum diffuse white W_{in}. They observe that the highlight regions contain more high spatial frequency content than diffuse image parts due to quick changes in the surface shading. They proposed a set of low-pass filters combined with morphological operations, which can automatically detect highlights (refer to Fig. 8.4). The Gilchrist theory of lightness perception [71] may provide some insight toward an automatic selection of W_{in} and W_{out} parameters. This theory relies on the notion of the reference white point, which is conceptually similar to the concept of W_{in}. The Gilchrist theory has already been employed for tone mapping [14], which is also based on linear contrast scaling within segmented image regions (frameworks) with clearly different luminance levels.

Banterle et al. [99] investigate nonlinear contrast scaling by inverting simple tone-mapping operators based on exponential and sigmoid functions. Visually the most compelling results have been obtained by inverting the photographic tone-mapping operator [65]. The

FIGURE 8.3: The view of an image with enhanced dynamic range as it would appear on an HDR display is simulated. The dynamic range has been enhanced using linear scaling (left) and the approach proposed in [98] with $W_{out} = 67\%$ (right). Because of dynamic range limitation on print only the appearance of the diffuse image part is simulated and the highlights can be properly seen only on an HDR display. The original image appearance as tone mapped for reproduction on an LDR display can be seen in Fig. 8.4. Note that the linear scaling may lead to overall too bright image appearance. Images courtesy of Laurence Meylan of General Electric, Scott Daly of Sharp Laboratories of America, and Sabine Süsstrunk of Ecole Polytechnique Fédérale de Lausanne (EPFL). Reprinted from [102] with permission from SPIE.

authors observed that when using this approach, they cannot expand the dynamic range to arbitrarily high values due to quantization errors manifesting in contouring artifacts in particular in bright image regions, in which the sigmoid function strongly compresses contrast. (The authors do not report any problem with saturated dark pixels.) To address the contouring problem they create an interpolation map, which is used to smooth shading of pixels that belong to the high luminance areas (refer to Fig. 8.5). The interpolation map is built in two steps. At first importance sampling over the pixel intensity distribution in the input LDR image is performed to find a set of virtual light sources that energy-wise represent the whole image and are concentrated mostly in high luminance regions (Fig. 8.5(center)). In the second step, density estimation over these light sources is performed for every pixel to obtain a smooth interpolation map (Fig. 8.5(right)). The interpolation map is finally used to blend between the original LDR image and the range-expanded image obtained though the sigmoid function inversion.

The authors validate their approach by comparing the reference HDR images against their range-expanded counterparts using the HDR VDP (refer to Section 4.2). A vast majority of perceivable differences reported by the metric come from the highlight and light source regions in which the luminance values of reconstructed pixels are selected in an ad hoc manner (refer to Fig. 8.6). This is a general problem for all discussed so far methods that focus on enhancing contrast and suppressing contouring artifacts, but do not pay much attention to clipped pixels both in dark and light image regions. In Section 8.4, we discuss solutions aimed at this problem.

FIGURE 8.4: Highlight and light source detection in LDR images using a segmentation algorithm as proposed in [102]. The detected highlight and self-luminous objects are marked in red. Images courtesy of Laurence Meylan of General Electric, Scott Daly of Sharp Laboratories of America, and Sabine Süsstrunk of Ecole Polytechnique Fédérale de Lausanne (EPFL). Reprinted from [102] with permission from SPIE.

In all LDR2HDR techniques discussed in this section, the goal of applying inverted tone mapping was to obtain visually plausible results. These methods do not give any insight what are the actual scene radiance values, which can be considered as an ultimate goal of any solid-grounded LDR2HDR reconstruction. In the following section we discuss techniques aiming at this goal.

8.3 SINGLE IMAGE-BASED CAMERA RESPONSE APPROXIMATION

The camera response function relates the scene luminance values to image pixel intensities captured in an image. Thus, if the inverse camera response function is known, the scene radiance map can easily be reconstructed. The problem of recovering the camera response function based on multiple, differently exposed images of the same mostly static scene is relatively well researched (refer to Section 3.2). A challenging question arises how to reconstruct the response function based on a single image without any knowledge of camera used for capturing,

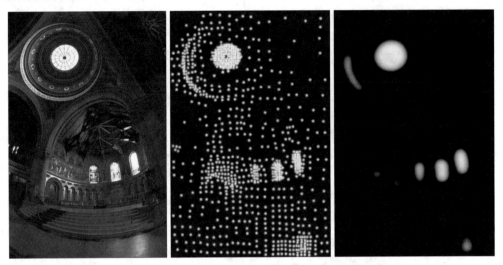

FIGURE 8.5: Interpolation map construction. Tone-mapped image (left) is importance sampled to a set of virtual light sources (center), which through density estimation process is converted in to the final interpolation map (right). Images created by Francesco Banterle. Copyright: Warwick Digital Laboratory, University of Warwick.

exposition parameters, and the captured scene characteristic? This is a typical situation for legacy images and video.

The camera response function should compensate for camera optic imperfections and sensor response nonlinearity, as well as image enhancement and tone mapping intentionally performed by camera firmware altogether. In many practical applications, the camera response function is often approximated by a simple gamma correction curve in which case some standard gamma value, e.g., 2.2 is usually assumed. Farid [103] proposes a more principled approach in which the gamma value can be blindly estimated in the absence of any camera calibration information based on the single image (the so-called blind inverse gamma correction). It turns out that gamma correction introduces to the image several new harmonics whose frequencies are correlated to the original harmonics in the image. There is also a strong dependence between the amplitudes of the original and newly created harmonics.

It can be shown that such higher order correlations in the frequency domain monotonically increase with increasing nonlinearity of gamma correction. Tools from the polyspectral analysis can be used to detect such correlations, and by searching for the inverse gamma, which minimizes such correlations, the actual gamma correction originally applied to the image can be found.

In practice, the gamma function is only a crude approximation of the camera response and by applying a simple inverse gamma correction to an image the accuracy of reconstructed

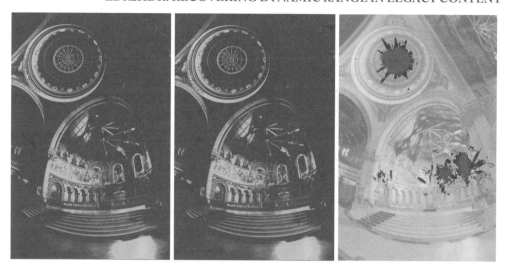

FIGURE 8.6: Radiance maps for the original HDR image (left) and its LDR image-based reconstruction (center). Pseudo-color encoding is used to depict radiance values with blue, green, and red roughly corresponding to 10, 100, and 700 cd/m^2. HDR VDP is used to predict perceivable differences (right) between the radiance maps shown in (left) and (center). In the perceivable difference map (right) red color denotes pixels for which the difference is over 1 just noticeable difference (JND) unit. Images created by Francesco Banterle. Copyright: Warwick Digital Laboratory, University of Warwick.

radiance map can be affected. Lin et al. [104] show that for a single LDR image the camera response curve can be more precisely reconstructed based on the distribution of color pixels in the proximity of object edges. The most reliable information for such reconstruction is provided by edges separating the scene regions of uniformly distributed and significantly different color (radiance values) R_1 and R_2 (refer to Fig. 8.7(a)). For a digitized image of the scene using a camera featuring the linear response, the color I_p of pixel representing precisely the edge location should be then a linear combination I_1 and I_2 (refer to Fig. 8.7(b)). The partial coverage of pixel area by each of the two regions decides about the contribution of I_1 and I_2 values into the pixel color I_p. However, due to the nonlinearity in the camera response the actual measured color M_p may be significantly different from such a linear combination of measured colors M_1 and M_2 (refer to Fig. 8.7(c)), which correspond to I_1 and I_2. By identifying a number of such $\langle M_1, M_2, M_P \rangle$ triples and based on the prior knowledge of typical real-world camera responses a Bayesian framework can be used to estimate the camera response function. By applying inverse of this function to each triple $\langle M_1, M_2, M_P \rangle$, the corresponding $\langle I_1, I_2, I_P \rangle$ should be obtained such that I_p should be a linear combination of I_1 and I_2. Applying such inverse response function to all image pixels results in reconstruction of the scene radiance map.

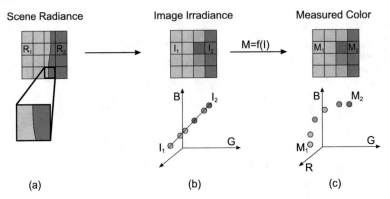

FIGURE 8.7: Color distortions in edge regions due to nonlinearity in the camera response. (a) Two regions in the scene, which are separated by an object edge, feature distinct spectral radiance R_1 and R_2 values. (b) Hypothetical linear image sensor maps R_1 and R_2 values into I_1 and I_2 values in the RGB color space. Due to the scene radiance digitization by the sensor, the color of each pixel on the edge is a linear combination of I_1 and I_2 with weights proportional to the covered area on the left and right sides of the edge. (c) A nonlinear camera response f warps these colors resulting in their nonlinear distribution. Redrawn from [104].

The authors observe that their method leads to a good accuracy in reconstruction the radiance map. The best accuracy is achieved when the selected edge color $\langle M_1, M_2, M_P \rangle$ triples cover a broader range of brightness values for each color channel. The method may not be very accurate for images that exhibit a limited range of colors. By using $\langle M_1, M_2, M_P \rangle$ triples from additional images captured with the same camera, the accuracy of the camera response reconstruction can be further improved. Obviously, the radiance information in saturated image regions cannot be recovered, and we address this problem in the following section.

8.4   RECOVERING CLIPPED PIXELS

Another problem with legacy images are image regions completely saturated due to intensity clipping of brightest and darkest images regions. The problem of lost information reconstruction is clearly under-constrained with a number of possible solutions that lead to the same appearance of an LDR image. Since under- and over-exposed image regions may contain only sparse information, learning approaches that rely on finding correspondences in a predefined database of LDR and HDR image pairs seems to be a very difficult task. The most promising results have been obtained so far using inpainting and texture synthesis techniques specialized in repairing damaged images or removing unwanted objects.

It can be observed that many LDR images, which are difficult cases for tone-mapping inversion approaches, may contain similar textures whose details remain intact in some image regions while they are clipped in very dark or bright image regions. Wang et al. [105] investigate texture transferring from such well-exposed regions by drawing from the texture synthesis literature. The authors call their approach *HDR hallucination*. The texture transfer in the LDR2HDR setting is actually more complex due to diversity of lighting conditions, which is usually not the case for traditional texture synthesis. To simplify this problem, the authors employ bilateral filtering to decompose inverse-gamma corrected LDR image (a roughly reconstructed radiance map of the scene) into a low-frequency illumination component and a high-frequency texture component [68]. Then saturated illumination component is reconstructed via interpolation from a linear combination of elliptical Gaussian kernels, which are fitted to non-saturated pixels around the over-exposed region. If needed, the fitted illumination function can be further manually adjusted. The high-frequency texture component is reconstructed via constrained texture synthesis [106] based on the source texture and destination location, which are manually indicated by the user. To correct for perspective shortening or properly align texture structure or semantic information the user draws a pair of strokes in the source texture and destination image region, and then the source texture is automatically warped to the required size and orientation. Poisson editing is performed [107] to smooth out transitions between the synthesized textures and the original image. Overall the proposed technique works remarkably well and its failure cases are mostly related to the lack of appropriate source textures in the image to be transferred. In such a case, another image can also be used to successfully transfer originally missing texture.

8.5   HANDLING VIDEO ON-THE-FLY

Rempel et al. [100] proposed on-the-fly solution to handle legacy video, which combines altogether all important elements of LDR2HDR processing such as reverse tone mapping, decontouring, contrast enhancement, and separate handling of highlight and light source regions. All these elements have been discussed in Sections 8.1–8.4, but in the proposed solution the emphasis is on its robustness (should not produce disturbing artifacts), automatic operation for preset parameters based on the HDR display characteristics, high computational performance and good temporal coherence of employed image processing algorithms.

Figure 8.8 shows the algorithm overview. At first an input LDR image is transformed from the ready-to-display, perceptually uniform, nonlinear representation (luma) to a linear space, which approximates luminance in the original space. For this purpose simple inverse gamma operation is performed and a gamma curve of 2.2 is used, which is standard in video

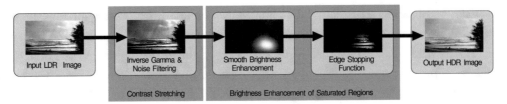

FIGURE 8.8: Overview of LDR2HDR processing. Image courtesy of Allan Rempel et al. [100]. © 2007 ACM, Inc. Used by permission.

and television formats. In the next step, contrast is stretched by simple mapping of linearized pixel values to absolute luminance values reproduced by the target HDR display. The authors limit contrast stretching to up to 5000:1, which always leads to improved image quality without causing artifacts that may arise for some images. Of course, even higher contrast stretching could lead to visually better results for some images, but the algorithm robustness and automatic operation requirements justify this hard limit on the maximum contrast. The contrast stretching may magnify noise and compression artifacts as well as may lead to visible contouring artifacts (refer to Section 8.1) in particular for poorer quality footage. In this case optionally bilateral filtering is performed, which is tuned to the possible artifact level while preventing blurring image features. Since filtering is performed in the perceptually non-uniform luminance domain the variance of the photometric Gaussian factor in the filter is adjusted for each luminance level to the quantization thresholds.

As found in [82, 81] to achieve good appearance of HDR images both luminance and brightness should be simultaneously increased. For this reason, in the next processing step, as shown in Fig. 8.8, smooth brightness enhancement is performed in the neighborhood of saturated image regions. At first such bright regions are identified by simple thresholding of pixels with RGB values over 230 (for video) and 254 (for photographs) in at least one color channel. The resulting bright pixel mask is strongly blurred with a filter, whose parameters are tuned to remove most of energy with spatial frequencies higher than 0.5 cpd from the mask signal (Fig. 8.9(upper-right) shows an example of brightness enhancement mask in red). For the remaining low spatial frequencies, the human eye is not very sensitive as predicted by the contrast sensitivity function (CSF), which effectively means that such smooth brightness enhancement proportional to the intensity of pixels in the mask should not introduce visible artifacts. An edge-stopping function is introduced to the blurred mask to prevent brightness enhancement in neighboring darker regions which are separated from the bright pixels by strong edges. An efficient implementation of mask blurring and edge-stopping filters are achieved using Gaussian image pyramids. Figure 8.9 shows the results obtained using this method.

FIGURE 8.9: On-the-fly video LDR2HDR processing: (upper-left) input LDR image, (upper-right) brightness enhancement mask, (lower-left) two virtual exposures of the reconstructed HDR image featuring contrast 9300:1, and (lower-right) the same HDR image shown on a Brightside DR37-P HDR display partially covered by a 10% neutral density filter to demonstrate details in bright image regions. Images courtesy of Allan Rempel et al. [100]. © 2007 ACM, Inc. Used by permission.

8.6   EXPLOITING IMAGE CAPTURING ARTIFACTS FOR UPGRADING DYNAMIC RANGE

Scattering of light inside the lens is very apparent in the capture of high dynamic range images, defining a limit to the dynamic range that can be captured with a camera [108]. Such scattering can be modeled with point spread functions (PSF) and removed using deconvolution [109]. However, precise estimation of the PSF is not trivial especially that its shape is non-uniform across the image. Deconvolution may also lead to high quantization noise in strongly veiled image regions, due to insufficient precision of real scene information. Recently, Talvala et al. [110] have demonstrated that by placing a structured occlusion mask between the scene and the camera, direct and indirect (scattered) light falling on the camera sensor can be separated. For a given position of the mask, the sensor elements, which are occluded by the mask, are illuminated by only scattered light. By jittering the mask position and capturing HDR images for each such position the amount of scattered light can be estimated for each pixel and then removed from the final HDR image. A practical problem with this technique is that the scene must be static, and the mask must be placed near the scene in order to be in camera focus so that its contribution to the intensity of non-occluded by the mask pixels is reduced.

8.7 CONCLUSIONS

As we discussed in Chapter 7 in coming years rapid development of HDR display technology can be anticipated. The process of standardization for lossy HDR image and video formats is just initiated (refer to Chapter 5), however, a number of years will be required before standards accepted by the industry will emerge. For this reason, the problem of legacy content enhancement is so urgent. In this respect, robust on-the-fly solutions as presented in Section 8.5 are of particular importance, since they can be embedded in the new generation of displays and tuned to obtain the best performance for a given display type. This solution enables to enjoy HDR content without waiting for painful format standardization and broadcasting HDR-enabled video signal. However, such dynamic range enhancement is ill-posed problem in which precise reconstruction of original HDR content is difficult and often not possible. For this reason, the development of algorithms enabling blind reconstruction of tone mapping will be important research topic in coming years. Also, robust detection of highlights and light sources in the original LDR footage and then restoration of missing information in saturated image regions is another challenge. In professional applications off-line dynamic range restoration, perhaps involving the user interaction for selected frames, and then propagation of restored information for the remaining frames, can be envisioned. In this chapter, we did not discuss the problem of color gamut enhancement, which will be important with constantly improving display technologies enabling wider gamuts, and thus enabling more saturated and vivid colors.

CHAPTER 9

HDRI in Computer Graphics

Recent developments in computer graphics and HDR imaging demonstrate strong mutual dependence. Computer graphics is a continuous source of an HDR image and video content. Synthetized HDR images are a natural outcome of more engineering oriented aspects of computer graphics such as physically-based image rendering, but recently also entertainment applications such as cinematography and computer games greatly benefit from more precise HDR pixel representations. On the other hand, the HDR images and video captured in the real-world are a precious source of the input data for image-based rendering and modeling in graphics. In this chapter, we discuss these important aspects of convergence between graphics and HDR imaging. To be aligned with the main topic of this book, whenever possible we focus on HDR video applications, but in some cases only static HDR images have been used so far. It can be envisioned that with quickly progressing HDR video cameras technology, image sequences will effectively replace static images in many of the discussed here applications.

9.1 COMPUTER GRAPHICS AS THE SOURCE OF HDR IMAGES AND VIDEO

At present, multi-exposure techniques and specialized HDR cameras are the main source of HDR images and video (refer to Chapter 3). While the multi-exposure techniques have been invented and applied first for the traditional film technology [12, 111], they gained real popularity when digital cameras with manually controlled exposures have been available. The development of full-fledged HDR cameras is just a matter of recent years. However, historically the first HDR images resembling photographs have been obtained in lighting engineering and realistic rendering communities. While these two communities have been working mostly independently aiming at different goals, they have had common interests in physically-based lighting simulation. Such simulation was always important in lighting engineering, but often limited to the estimation of numerical values of illumination at selected points in the designed environments, e.g., workspace. The progress made in the meantime in graphics had a significant impact on lighting engineers and designers, who showed more and more interest in realistic image synthesis as well. In particular, the work of Greg Ward and his publicly available

RADIANCE system [112, 113] popularized image synthesis in the lighting community. In computer graphics, image synthesis has always been one of the major goals, but just in the mid-eighties researchers started to combine realistic image synthesis with physically-based lighting simulation [114, 115, 116]. The first inspiration on how to deal with this problem came to graphics from the heat transfer literature (mostly finite element methods [117]) and lighting engineering (also Monte Carlo methods [118, 119]).

Physically-based lighting simulation required valid input data expressed in radiometric or photometric units. It was relatively easy to acquire such data describing light sources, because high-profile manufacturers of lighting equipment measured, and often made available directional emissive characteristics of their luminaires (the so-called goniometric diagrams). It was far more difficult to obtain valid reflectance characteristics of materials (the so-called bi-directional reflectance distribution function—BRDF). However, the assumption of Lambertian (perfectly diffuse) reflectance model has been predominant at early days of lighting engineering and realistic graphics, which greatly simplified the computation. It was relatively easy to estimate the surface albedo (a single scalar value), which fully characterizes the reflectance for Lambertian surfaces. In the nineties, lighting simulation methods progressed to handle more general environments efficiently, and more advanced BRDFs have been measured (refer to Section 9.2.2) or expressed using physically-valid analytical models.

Physically-based lighting simulation with the use of physically-valid data, which describe the rendered scenes, resulted in a good approximation of illumination distribution with respect to the corresponding real-world environments. Also, pixels in rendered images were naturally expressed in terms of radiance or luminance values, which is the distinct characteristic of HDR images. To store such images efficiently first compact HDR image formats have been developed, such as the RGBE format (refer to Section 5.1.2) proposed by Ward as a part of his RADIANCE package [49]. Also, early tone-mapping techniques appeared to enable image display on devices with a limited dynamic range [120, 121, 122]. Figure 9.1(left) shows a typical example of realistic image rendered using Monte Carlo methods. Figure 9.1(right) shows the corresponding HDR image that was captured in the actual real-world scene.

While realistic rendering software is a source of HDR images and video for almost two decades, recently available graphics processing units (GPU) and major game consoles upgraded their rendering pipelines to the floating point precision, which effectively enabled HDR image rendering. Thus, in the years to come computer games and other real-time applications running on these platforms will be an important source of HDR image sequences. In the simplest case, such HDR-enabled platforms could be directly connected to an HDR display offering even more immersive game experience. In fact, due to lack of standardization such a direct connection, could require some engineering efforts to accommodate specific signal requirements e.g., Brightside DR37-P requires special steering of an LED array in its backlit

FIGURE 9.1: Atrium of the University of Aizu: (left) rendered image, and (right) HDR photograph. Accompanying web page http://www.mpi-inf.mpg.de/resources/atrium/ provides with the complete data set required to render this image. Also, the results of lighting simulation have been compared to the measurement data in the actual scene.

device [2]. This problem has been successfully solved by Ghosh et al. [123], who additionally enhanced the immersive game experience by adding surround lighting that can be seen be in the user's peripheral field of view. However, even without having the access to an HDR display, computer games benefit greatly from many HDR visual effects which are difficult to model convincingly using LDR game pipelines:

- Glare (dazzling) effects around strong lights and bright highlights, which are modeled using an image-processing approach by applying a pyramid of carefully tuned low-pass filters with different spatial support to every bright pixel. The filter pyramid effectively spreads bright pixel intensity in the neighborhood causing characteristic blooming pattern, which reduces contrast and thus detail visibility in the proximity of bright image regions. An alternative, cheaper, but less general way of glare modeling is to impose sprites (pre-computed bit maps with the bloom pattern) around strong light sources (light reflections are more difficult to handle). The sprites apart from the

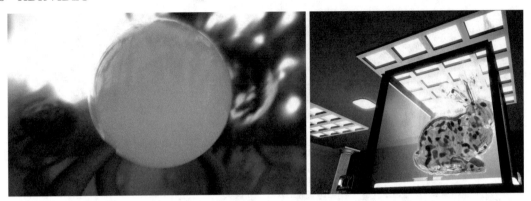

FIGURE 9.2: Real-time GPU rendering with HDR effects. (left) Realistic reflection with Fresnel's effects on the surface of the white plastic ball. Visually correct motion blur, depth of field, and glare effects. (right) Interesting volumetric refraction and reflection effects inside the foggy box. All computation performed for HDR pixel intensities. Real-time tone mapping applied prior to the image display. Images courtesy of Ivo Ihrke, Gernot Ziegler, Art Tevs, Christian Theobalt, Hans-Peter Seidel of MPI Informatik and Marcus Magnor of Technische Universität Braunschweig.

blooming appearance can add camera-triggered effects such light streaks (stars) caused by diffraction over the diaphragm blades and ghosts caused by internal reflection in the multiple-lens optical system. The sprites are feasible only for a small number of bright, regularly shaped, and small light sources such as the sun, car headlights, lamps, and so on.

- Exposure control with dark and light adaptation using a tone-mapping technique for dynamic sequences.
- Depth-of-field effect with the shape of aperture stop.
- Motion blur performed for HDR pixels to avoid intensity clamping typical for LDR approaches.
- Bright reflections (refractions) of strong light sources (e.g., the sun) in the surfaces of low reflectance (transmission), which cannot be reproduced in the LDR setting due to light intensity clamping.

Figure 9.2 presents some of discussed effects as rendered on a modern GPU with real-time performance. In game applications, the main goal of modeling these effects is to improve the visual attractiveness and believability of images, while their physical correctness is of secondary importance. Some of the discussed effects such as glare are presented in the computer graphic literature as an integral part of widely understood tone mapping (refer to Chapter 6).

9.2 HDR IMAGES AND VIDEO AS THE INPUT DATA FOR COMPUTER GRAPHICS

Machine vision and computer graphics are rapidly converging disciplines. Image-based rendering is a prominent example of such convergence, where computer graphics techniques enable changes in the scene viewpoint, lighting conditions, object appearance, or even image content. When video is used as an input, all these modifications can be performed in temporal domain, and additionally kinematics and dynamics can be manipulated. Image-based rendering still outperforms traditional 3D rendering in terms of achieved realism, and obviously acquiring images is much easier than building scene model using standard 3D graphics tools. Vision and graphics coupling is even more obvious in augmented reality applications in which real-world images and video must be seamlessly mixed with rendered objects. Finally, image-based modeling can be used for efficient acquisition of data required in graphics such as 3D scene geometry or material reflectance characteristics.

HDRI technology has great potential in all discussed image-based techniques used in graphics, because it is less sensitive under extreme lighting conditions. This means that virtually all pixels convey potentially useful information, while using traditional cameras such information can be lost in under- and over-exposed image regions. The HDR camera has also great potential as a radiometrically (photometrically) correct measurement device (refer to Section 3.2), which in single image provides millions of independent measurements acquired at once for all pixels. Such physical correctness is required in particular in realistic image synthesis, which is one of the mainstream applications in 3D graphics. In the following sections, we present applications of HDR imaging for acquisition of scene lighting and surface reflectance, which greatly contribute to the final appearance of rendered objects.

9.2.1 HDR Video-Based Lighting

Traditionally, in realistic 3D image synthesis lighting is modeled by specifying a certain number of directional, point, or area (usually rectangular or circular shape is assumed) light sources distributed in the scene. In physically-based rendering, the computation of interreflection is additionally performed to account for indirect lighting illuminating the scene. In cinematography more control over lighting distribution may be required for artistic reasons, and indirect lighting is often replaced by inserting into the scene a huge number of individually-controlled local lights. Another important reason for such a non-physical approach is huge costs to compute the interreflection given the complexity of scenes in modern computer-generated movies. Only recently one bounce of indirect lighting has been used in high profile productions like Shrek 2. Game industry relies mostly on direct lighting and lack of interreflection compensates using ambient lighting, which in more advanced cases may be modulated based on purely geometrical

FIGURE 9.3: Realistic rendering of the dragon model with measured bi-directional texture function (BTF) of leather [124]. Captured real-world lighting, which is visible at background, is used to illuminate the model. Image courtesy of Gero Müller, Ralf Sarlette, and Reinhard Klein of the University of Bonn.

visibility considerations (the so-called ambient occlusion technique). However, in all discussed cases resulting images have usually a synthetic look, which can be easily distinguished from photographs. The exception are cinematographic applications in which more realistic effects are achieved through time consuming tweaking of local lighting parameters.

Much better realism can be achieved when a synthetic 3D scene model is illuminated by camera-captured real-world lighting (refer to Fig. 9.3). The technique is called *image-based lighting* (IBL), and the problem of costly interreflection computation is less-pronounced for this technique since images capture both direct and indirect lighting simultaneously. The only problem is to account for interreflection between the illuminated object and the scene, but this is often negligible, e.g., in the game scenario moving characters usually do not contribute much into indirect lighting of the whole scene. However, what makes the IBL so compelling comes from the side of human visual system (HVS), which is strongly adapted to real-world lighting conditions and makes many implicit assumptions about statistical regularities in such a lighting [125]. The geometrical structure and other statistics of real-world lights are often needed to disambiguate information about surrounding objects. Note that the same amount of light may fall onto the human eye retina when reflected from strongly illuminated surfaces that are poor reflectors and identically-shaped surfaces that are good light reflectors located in a dim environment. The human visual system can easily distinguish both situations by discounting the illuminants, which computationally is an ill-posed problem of lightness determination that requires some assumptions about the scene lighting to be solved [126, 127]. Through psychophysical experiments with computer-generated images, Fleming et al. [125] have shown that the human observer ability to notice even subtle differences in the material appearance (surface reflectance characteristics) is much better under real-world lighting conditions than commonly used point light sources. Realistic lighting improves also the ability to discriminate between rendered objects, whose shape is only slightly different [128]. This observation has strong implications in the industrial design practices, and for example images of new car models used for advertisement purposes are predominantly rendered as illuminated by captured real-world lighting (e.g., using the SpheroCam HDR camera [129]). On the other hand, the HVS sensitivity to the differences in reflectance properties strongly depends on the object shape [130].

Clearly, real-world lighting is desirable in many engineering applications and would improve the believability of virtual reality systems notoriously lacking realism in rendering. Real-world lighting is indispensable in many mixed reality applications, in which virtual objects should be seamlessly merged with a real-world scene [131].

Traditionally, real-world lighting is captured into *the environment map* (EM), which represents distant illumination incoming to a point from thousands or even millions of directions that are distributed over a sphere (hemisphere). HDR technology is required for the environment map acquisition to accommodate high contrasts in the real-world lighting. Under static conditions low dynamic range cameras and a multi-exposure technique can be used to capture two HDR images, which fully cover a spherically-shaped mirror light probe [12]. For dynamic light capture an HDR video camera with a fish-eye lens is the best solution to obtain hemispherical environment map, which we call the video environment maps (VEM). Existing multi-exposure techniques for video are limited just to two exposures [15], which may not offer sufficient dynamic range for robust capturing of high contrast lighting.

An important question concerning visually tolerable distortions in captured HDR EM and VEM arises due to the limitations in camera resolution and geometry distortions introduced by a fish eye lens. Ramanarayanan et al. [132] conducted a psychophysical study in which they investigated the impact of these two factors on the visual equivalence in object material and shape perception. It turned out that even significant amount of blur in EM lighting still leads to visually equivalent images, in particular for less glossy objects, which act as low-pass filters for reflected lighting [133]. Lighting geometry distortions may be more objectionable, which means that stronger warps of environment maps can be wrongly interpreted as change in the object shape. In this case, the HVS would expect that perceivable distortions in the EM reflection come rather from imperfections in the object surface than deformed shapes of light sources, which is less likely scenario in the real-world environments. However, these problems are negligible for lens distortions and image resolutions offered by existing HDR video cameras.

Two rendering techniques: precomputed radiance transfer (PRT) and environment map importance sampling are prevailing solutions in interactive rendering with EM lighting. Both techniques naturally support rotations of EM and can easily be extended to handle VEM. We briefly characterize these techniques and then we focus on their successful applications with the use of VEM. For more general discussion of IBL techniques, which concerns mostly static lighting, please refer to an excellent survey in [6] Chapter 9.

Video Environment Maps in Precomputed Radiance Transfer

Interactive rendering of realistic objects illuminated by large light sources is a difficult problem, in particular, if such light transport effects as shadows, interreflections, and sub-surface scattering are taken into account. For scenes that are illuminated by the EM the most costly computation comes from testing visibility and integrating incoming lighting over all hemispherical directions (spherical for non-opaque objects). PRT techniques relegate these costly computation to preprocessing, which dramatically reduces the computation load at the rendering stage [134].

Essentially PRT computes the illumination of each point in the scene (often mesh vertices are only considered) as a linear combination of incident lighting, which may come from all directions over the sphere, but at the same time it is assumed that light source (environment map) is far away from the scene. A direct consequence of this assumption is that for all non-occluded points in the scene, the same incident lighting always comes from a given direction, which greatly simplifies the computation and bookkeeping of incident lighting. This is also a realistic assumption for outdoor scenes illuminated by sky lighting, but may fail for some indoor scenes with spatially varying direct lighting (at the end of this section we discuss how to overcome this limitation).

To encode incoming lighting an efficient spherical basis function such as *spherical harmonics* (SH) is commonly used in PRT techniques. The SH basis has a very powerful property: The integral over a product of two spherical functions reduces to the dot product of the SH coefficients of these two functions. Let us recall that the global illumination problem is essentially equivalent to the solution of such an integral, but for the product of three functions: reflectance (BRDF), visibility, and incoming lighting [135]. For this reason, the reflectance (for Lambertian surfaces just a scalar value) and visibility information is usually concatenated into a single function called the transfer function. The transfer function encapsulates the whole light transport information from the directional light sources (represented by pixels in the EM) to each point in the scene, and it is computed at the pre-processing stage and stored as SH coefficients. The transfer function includes the direct EM visibility/occlusion information for each point in the scene, as well as directional visibility and energy attenuation information for indirect light transport. The lighting function is projected on the SH basis functions on-the-fly for each VEM frame, which enables dynamic lighting simulation. Such a projection is very fast and can be easily done at interactive speeds (e.g., for the VGA-resolution video of 640×480 pixels per frame).

Lighting and transfer functions for Lambertian surfaces are usually projected into 25 SH basis functions for each sample point. In general, this leads to good visual results, but only slowly changing and smooth lighting can be reproduced, e.g., soft shadows. Thus, lighting details that require high spatial frequency patterns, cannot be reproduced, e.g., sharp shadow boundaries. For more general reflectance functions (BRDF) for which the incoming lighting directions are important, a matrix of spherical harmonic coefficients with the transfer vectors for each of those directions must be considered. In practice, matrices of 25×25 coefficients are commonly used [134]. Since the transfer vectors (matrices) are stored densely over the scene surfaces (usually for each mesh vertex) an important issue is the data compression, which can be efficiently performed using standard tools such as principal component analysis (PCA) and clustering [136]. Recently, the limitation of low-frequency lighting, which is inherent for the SH basis, has been lifted using the wavelet basis functions [135]. Using the approach proposed by Ng et al. both soft and sharp shadows can be rendered, but very dense mesh is required to reconstruct the lighting function precisely and it is not clear how to include interreflections into this framework.

Another serious drawback of PRT techniques is the assumption that the scene is static for the transfer function computation at the preprocessing stage. If the interreflection computation is not required, this assumption can be relaxed using the SH exponentiation approach [137], which can efficiently handle soft shadows for deformable objects. However, PRT techniques are useful in many technical applications in which scenes with static geometry and dynamic lighting are considered and global illumination at interactive rates is important.

We present an example of such an application in which PRT techniques have been used in a virtual reality (VR) system aimed at simulation of lighting in the car interior. The interior can be illuminated by VEMs that have been captured under various driving conditions and are visible through the car windows. Figure 9.4 shows the acquisition system mounted on the roof of a car, which is composed from two HDR video cameras with fish-eye lenses for the windshield view and sky lighting capturing. The main goal of the VR system is to study the impact of such dynamic real-world lighting, which is captured for the actual driving conditions, on the visibility of information displayed on the LCD panel mounted in the car cockpit. This application scenario is similar to the simulation of free driving in an environment in which buildings, trees, and other occluders change the amount lighting penetrating the car interior. This requires that a global illumination solution responds interactively to lighting changes for an arbitrary position of the driver head (virtual camera position), which can be easily achieved using PRT. Figure 9.5(left) shows a snapshot of interactive PRT rendering. Figure 9.5(right) shows

FIGURE 9.4: HDR video environment maps (VEM) acquisition system equipped with two photometrically-calibrated HDRC VGAx (IMS CHIPS) cameras for the sky lighting and windshield view capturing.

the result of off-line rendering using a more precise path-tracing method, which also employs the captured VEM to model input lighting. To improve the immersion experience the CAVE environment with five stereo-projected screens is used for displaying the car interior. Also, a head tracking system is employed to monitor the driver's head position, which is important to properly warp the car interior images projected on the CAVE screens. The head-tracking system enables also to model light reflections in the LCD panel as seen from the drivers' point of view.

Figure 9.6 shows the appearance of LCD panel under the global illumination conditions for dynamic VEM lighting as displayed on an HDR monitor. All rendered images are inherently HDR because physically-correct car model and calibrated VEM lighting have been used for the global illumination computation, which is performed with the floating-point precision. Since the dynamic range of an HDR monitor is significantly higher than that of a typical LCD panel that is mounted in the car cockpit, the visibility of information displayed for the driver can be tested under many external lighting conditions. Through the calibration of the HDR display the real-world luminance values can be reproduced for the LCD panel by taking into account both the panel emissivity as well as reflected lighting resulting from the global illumination simulation.

FIGURE 9.5: Snapshots of the car interior (left) rendered at interactive speeds using PRT techniques, and (right) computed off-line using the physically-accurate path tracing algorithm. Calibrated HDR VEMs have been used to model input lighting. Notice the cockpit reflections in the windshield for the path-tracing image. Images courtesy of Tom Annen of MPI Informatik.

Importance Sampling for Video Environment Maps

Many practical rendering algorithms achieve the best performance for very simple directional and point light sources. Such types of light are well suited for the shadow computation and shading using graphics hardware and ray tracing. In fact, more advanced area light sources are usually decomposed into a set of such simple lights. The same approach can be applied for the EM lighting, which is decomposed into a set of representative directional light sources due to the infinite light source distance assumption. Such a set should be equivalent to the source EM in terms of lighting energy, but also resulting shadows should be visually equivalent to the outcome of brute force integration of incoming lighting over all pixels in the EM. The human perception helps to achieve the latter goal, because under typical display observation conditions

FIGURE 9.6: The LCD panel appearance as a result of the global illumination computation for VEM lighting: (left) full global illumination, (center) display emitted light only, and (right) reflected light. To compute the reflected light, BRDF-driven importance sampling and PRT lighting querying has been performed. Images courtesy of Tom Annen of MPI Informatik.

it is safe to assume that the just discriminable change in contrast must be over 1%. In case all directional lights carry the same energy, which is the optimal condition in terms of image variance (noise) reduction, having more than 100 light sources that illuminate each point in the scene, makes the influence of each light undiscriminable. This leads to smooth shading without banding or contouring artifacts. The practical number of light sources to achieve this goal is roughly 200–300 because some lights can be occluded, and then the relative contribution of each non-occluded light sources could be greater than the discriminability threshold. This larger number of light sources is also required because the full EM contains all possible directions over the sphere, and each point in the scene, which represents an opaque surface, can be illuminated only by lighting coming from the upper hemisphere with the pole determined by the normal vector direction.

A number of techniques for the EM decomposition into visually equivalent set of directional light sources have been developed in recent years. However, a vast majority of these techniques have been designed for static EM, and they do not generalize well for the VEM case. The main problem is the computation performance, which is far from interactive and precludes the VEM frames processing on-the-fly directly for captured light. Another serious problem is lack of temporal coherence, which means that significantly different set of directional light can be selected even for moderate and local changes between the VEM frames. This results in severe flickering artifacts that are not acceptable.

Havran et al. [138] have proposed an algorithm specifically designed for on-the-fly VEM processing. To reduce temporal flickering, they use the same set of initial samples over the unit 2D square for each VEM frame. The samples are generated using the quasi-random 2D Halton sequence, which means that they are well stratified over the unit square surface. The Halton sampling enables adding new samples without affecting the position of existing samples, while good sample stratification properties are always preserved. This is important for the progressive image quality refinement and maintaining constant frame rate by adjusting the number of directional lights on-the-fly. In order to improve light sampling properties Lloyd's relaxation over the initial sample positions is performed at the pre-processing stage, which results in the blue noise properties of the sampling pattern [139, 140]. Figure 9.7(left) illustrates the resulting position of samples as mapped from the unit square to the hemi-sphere, which would be close to the optimal sampling pattern in terms of visible noise reduction for the uniform energy EM. In practice, the position of directional light sources is adjusted accordingly to the local energy distribution in the EM. As shown in Fig. 9.7(right) the directional lights are more densely concentrated in brighter EM regions, in particular around the sun location, while darker regions are represented only sparsely. This is achieved using the importance sampling procedure, which is well established in the Monte Carlo literature [141]. The pixel luminance values in the EM are treated as a discrete 2D probability density function (PDF). Then stratified Halton samples are transformed to samples drawn from the discrete PDF and mapped to spherical coordinates.

FIGURE 9.7: Distribution of samples for uniform intensity (left) and real-world captured (right) environment maps. The left image demonstrates a good stratification and blue noise properties of the initial sample distribution. These properties are partially maintained in the distribution of samples in the right image, which is a warped version of the sample positions in the left image. The importance sampling applied to the samples in the right image prevents folding and preserves neighborhood relations between samples as imposed by their initial position in the left image.

This procedure is described in detail in [142]. In fact, Havran et al. used slightly more involved sample transform method [143], which exhibits unique continuity and uniformity properties. The method guarantees the bi-continuity property for any non-negative PDF, which means that a small change in the input sample position over the unit square is always transformed into a small change in the resulting position of light source over the EM hemisphere. This property greatly improves temporal coherence.

Havran et al. [138] have built a complete system, which enables the HDR VEM acquisition and rendering with captured lighting at interactive speeds (refer to Fig. 9.8). A photometrically-calibrated HDRC VGAx (IMS CHIPS) camera with a fish-eye lens is used for the VEM acquisitions [144]. The inverse camera response (refer to Section 3.2) is used to transform captured RGB values into the luminance map. This luminance map is submitted to the importance sampling procedure to reconstruct a representative set of directional light source. Since even local changes in the VEM frame lead to global changes of the PDF, the direction of virtually all light sources may change from frame to frame, which causes unpleasant flickering in the rendered images. Havran et al. apply a perception-inspired, low-pass FIR filtering to the trajectory of each light motion over the hemisphere as a function of time. Since the energy in the environment map can fluctuate, in particular for scenes with incandescent lighting, filtering over all environment map energy is performed as well. A better stabilization of temporal artifacts is achieved, when a certain number of frames from the future are considered. For this reason, a delay of four VEM frames is introduced, which is essentially not objectionable because frame grabbing in their system works asynchronously in respect to

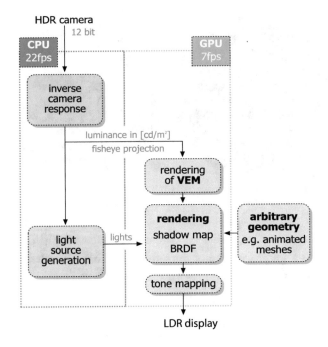

FIGURE 9.8: The HDR video capture and rendering system pipeline illustrating the distribution of tasks between CPU and GPU.

usually slower rendering. All computation discussed so far is performed on the CPU side of their system.

The GPU part is responsible for rendering. The luminance map acquired by the camera is at first displayed as the background and then all objects in the scene are rendered. Directional lights decomposed in CPU from each HDR VEM frame along with the shadow-mapping technique are used to illuminate the scene. The stratification and progressiveness properties of the Halton sequence permit adding more lights for selected angular regions in the EM without affecting the directions of already distributed lights. The directional light sources, which represent strong emitters such as the sun, can be clustered to reduce the cost of computing shadows. Finally, rendered GPU frames are tone mapped (refer to Section 6.1) and displayed.

The system presented by Havran et al. does not require any costly preprocessing, can handle fully dynamic geometry and arbitrary reflectance models evaluated on a GPU (refer to Fig. 9.9). The system does not support interreflection, but it seems that the instant global illumination algorithm [145] fits well its architecture. The main use of the proposed system can be envisioned in augmented reality applications in which real and synthetic objects are illuminated by consistent lighting at interactive frame rates (refer to Fig. 9.10).

FIGURE 9.9: A snapshot obtained using the Havran et al. system. Left: distribution of directional lights (marked as the green dots) over a VEM frame as captured using the fish-eye lens (top) and shown in polar projection (bottom). Right: Stanford BUNNY illuminated by the 72 directional lights.

Grosch et al. [146] have built such an augmented reality system capable of the diffuse interreflection computation (refer to Fig. 9.11). As in [138] an HDR video camera is used to capture dynamic lighting and at the same time another HDR video camera captures the scene view. The latter view is augmented in real time by adding virtual objects, which are illuminated by direct and indirect lighting components from the real scene (the influence of the

FIGURE 9.10: Comparison of the fidelity in the shadow and lighting reconstruction for the real-world and synthetic angel statuette illuminated by dynamic lighting. Real-world lighting is captured by the HDR video camera located in the front of the round table with an angel statuette placed atop (the right image side). The captured lighting is used to illuminate the synthetic model of the angel statuette shown in the display (the left image side).

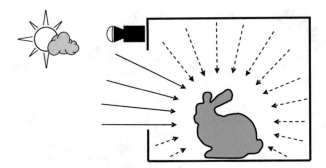

FIGURE 9.11: Virtual Bunny illuminated by daylight as captured by an external HDR video camera and indirect lighting simulated for the Cornell Box interior. It is assumed that the interior geometry and its reflectance properties are known to perform such a simulation. Image courtesy of Thorsten Grosch. Copyright the University of Koblenz-Landau.

virtual objects on the scene illumination is ignored). Direct lighting is computed using importance sampling of VEM, which additionally takes into account the visibility of virtual objects. Figure 9.12 summarizes the indirect lighting computation, which is performed for digitized geometry and material reflectance properties of the real-world scene. The hemisphere with captured lighting (effectively VEM as in [138]) is decomposed into a number of angular sectors and for each such a sector a basis irradiance volume (i.e., directional distribution of incoming lighting at the nodes of a uniform grid in the scene [147]) is pre-computed using the radiosity method. To find the actual indirect lighting at a given node, contributions from all basis irradiance volumes are re-scaled based on the captured VEM lighting and summed up at interactive speeds. The indirect lighting at any point at the virtual object is trilinearly interpolated based on illumination stored for neighboring nodes. Figure 9.13 shows the comparison of the Cornell Box scene augmented with the virtual teapot with respect to the ground-truth real-world view with the teapot obtained using a 3D printer. As can be seen, the system proposed by Grosch et al. can faithfully model virtual objects illuminated by distant direct and spatially varying indirect lighting at interactive framerates.

Wan et al. [148] proposed another algorithm suitable to handle VEM. They introduce a quad-tree over the sphere based on the adaptive subdivision of spherical quadrilaterals, which they call the Q^2-tree structure. They adaptively sample the EM based on an importance metric, which leads to finer quad-tree subdivision in brighter EM regions (refer to Fig. 9.14). A directional light source is created for every quad (stratum), which is a leaf-node in the Q^2-tree. The radiance emitted by all pixels in a given quad is summed and assigned to the corresponding light source, whose direction is jittered with respect to the quad centroid. To maintain temporal coherence, the authors adjust the Q^2-tree from the previous frame by

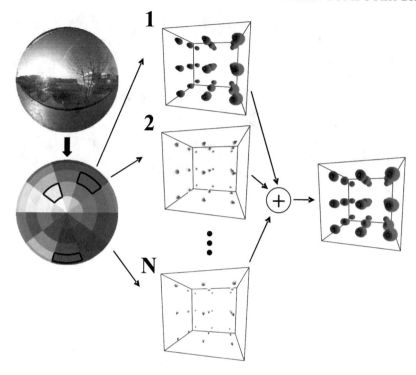

FIGURE 9.12: Combining basis illumination. For each solid angle sector of the fisheye lens an irradiance volume basis is computed using the radiosity method. The directional illumination distribution is computed for each node of the uniform grid and compressed using spherical harmonics for more efficient storage and access. The actual indirect lighting is obtained by combining basis illumination scaled by the actual lighting captured for each sector of the fisheye lens. Image courtesy of Thorsten Grosch. Copyright the University of Koblenz-Landau.

splitting leaf-node quads, which gain radiance (thus importance) with respect to the previous frame. Analogously, leaf-node quads are merged in the regions that become darker. A certain number of such merge-and-split iterations are considered, so that lighting changes are mirrored by the current Q^2-tree structure, and at the same time coherence with the previous frames is preserved whenever possible. The number of iterations decides whether lighting represented by Q^2-tree is more up-to-date, or more temporally coherent. The authors show that using their approach low-discrepancy sampling patterns are generated.

The distant lighting assumption inherent for traditional EM lighting holds well for many outdoor situations, but often fails for indoor scenes. In the latter case, when captured lighting is represented just by a single environment map and decomposed into a set of directional lights, the appearance of shadows cast by dynamic objects may look unrealistic. In environments

FIGURE 9.13: Comparison of virtual teapot appearance (left) as illuminated by captured direct and simulated indirect lighting with respect to its real-world counterpart printed using a 3D printer (right). Image courtesy of Thorsten Grosch. Copyright the University of Koblenz-Landau.

with dominant directional lights the resulting shadows are casted always in the same direction irrespectively of the dynamic object position. This problem can be significantly ameliorated when the directional lights are replaced with a representative set of point light sources with their fixed position in the scene. Korn et al. [149] have built an augmented reality system aimed toward achieving such goal[1]. In their system, they use two photometrically-calibrated HDRC VGAx (IMS CHIPS) cameras with fisheye lenses [144] as shown in Fig. 9.15. The HDR cameras are attached to the table and their fisheye lenses are upward directed. Additionally, a webcam camera can be seen, which is directed towards a marker on the table, where a virtual object is to be placed. The display shown in Fig. 9.15 presents the EM images captured by the cameras as well as the view from the webcam augmented with a virtual object, which is illuminated by captured lighting.

FIGURE 9.14: Adaptive sphere subdivision using the Q^2-tree technique. Redrawn from [148].

---

[1] Good background information concerning the lighting reconstruction using a stereo-camera pair, and then the scene augmentation with virtual objects is provided by Sato et al. [150]. However, their system is off-line and only static scenes are considered.

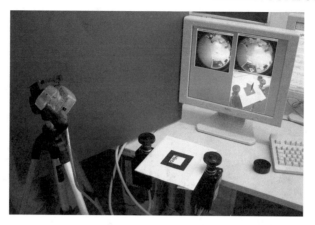

FIGURE 9.15: Light capturing system with stereo HDR video cameras (attached to the table). The webcam captures the scene view with the marker, which tracks position of the virtual object. The resulting augmented scene is shown on the display as well as two environment maps used to illuminate the scene. Image courtesy of Thorsten Grosch. Copyright the University of Koblenz-Landau.

The decomposition of VEM captured by the two video cameras into a set of point light sources is performed as follows. In one of the captured EM bright pixels are selected, and their corresponding positions in the second EM are found using the epipolar geometry. This narrows the search space to pixels located along the corresponding epipolar line. In fact, the epipolar lines are distorted into curves due to the image geometry imposed by the fisheye lenses. In practice, Korn et al. precomputed 500 epipolar curves and stored them in a look-up table to improve the correspondence search efficiency. When the corresponding light sources are found in the captured EM, then based on the known camera parameters and the distance between the cameras, 3D light source positions can be derived by means of simple triangulation. The light positions are tracked from frame to frame and updated along with changes in lighting as captured by the cameras. Figure 9.16 shows the real-world scene and the corresponding augmented scene with added a virtual box. Note a good match of shadows, which is achieved automatically due to real-world lighting capture.

9.2.2   HDR Imaging in Reflectance Measurements

High-quality modeling of surface reflectance properties contributes greatly to the realistic appearance of rendered objects. At present, analytical reflectance models are still predominant in low-end applications due to their compactness, but their use is often difficult due to non-intuitive and perceptually non-uniformly scaled parameters [151], which often do not have any physical meaning and cannot be measured for real-world materials. Also, the class of real-world

FIGURE 9.16: Real-world scene (left) and its augmented counterpart (right). The box floating over the table is a virtual object, which augments the video stream captured by the webcam (refer to Fig. 9.15). The box is illuminated by lighting captured by the two HDR cameras. Images courtesy of Thorsten Grosch. Copyright the University of Koblenz-Landau.

materials that can be convincingly represented using the analytical reflectance models is limited. For this reasons, many industrial and cinematographic applications, which require high fidelity or at least plausibility in the appearance of complex materials, relies on measured bi-directional reflectance distribution function (BRDF).

Bi-Directional Reflectance Distribution Function Acquisition
The BRDF is a 4D function, which is defined as the ratio of radiance outgoing in the direction (θ_o, ϕ_o) to irradiance (the radiant power per unit area) incident onto a material sample from the direction (θ_i, ϕ_i). For opaque surfaces, the BRDF is measured for all combinations of incoming and outgoing light directions over the hemisphere. Specialized gonioreflectometers with robotically controlled positions of the light source and detector with respect to the flat material sample are used for the high-quality BRDF measurement. Such a measurement can be performed much faster using a calibrated camera, which captures a curved material sample [152, 153]. In this case each pixel, which represents the material sample, effectively provides measurement data. Instead of capturing the spherical material probe, the appearance of real-world curved objects with spatially varying BRDF can be captured using a relatively small number of HDR images [154]. Sparse BRDF sampling over (θ_i, ϕ_i) and (θ_o, ϕ_o) direction pairs for each point on the object surface is compensated by exploiting the spatial coherence of BRDF for neighboring regions, and by fitting the measured data to an analytical reflectance model whose parameters change over the object surface. Figure 9.17 shows the acquisition setup used by Lensch et al. Figure 9.18 presents an object, whose geometry and spatially varying reflectance has been captured, as it is rendered under arbitrary lighting conditions.

FIGURE 9.17: Photograph of a setup used for capturing spatially varying BRDF [154]. In a photo studio covered with dark felt the following setup elements can be seen (from left to right): an HMI metal halide bulb serving as a point light source, metal spheres whose highlight configuration serves to track the light source position, object whose BRDF is acquired, and a Kodak DCS 560 camera used for multi-exposure HDR images acquisition. Image courtesy of Hendrik P. A. Lensch, Jan Kautz, Michael Goesele, and Hans-Peter Seidel of MPI Informatik and Wolfgang Heidrich of the University of British Columbia. © 2003 ACM, Inc. Used by permission.

Bi-Directional Texture Function Acquisition

All BRDF measurement techniques discussed so far are suitable for materials, which do not exhibit complex spatial structure. While such structure must be rendered to convey the material look-and-feel, it is usually impractical to include such fine scale details into the geometrical model. Also, complex light interactions within the fine structure due to light sub-surface scattering and self-shadowing cannot be captured by global illumination simulation due to excessive costs. These effects can be captured in the bi-directional texture function (BTF), which is a 6D texture representation that generalizes the BRDF by adding information on the sample point 2D position (u, v) over the surface A. Effectively each BTF sample is parametrized by its position (u, v) at A, and the incoming and outgoing light directions (θ_i, ϕ_i) and (θ_o, ϕ_o). Figure 9.3 shows an example of realistic rendering of dragon model covered with a leather BTF and illuminated by a captured HDR EM. Figure 9.19 shows a BTF measurement setup, in which a CCD camera is used to capture the material sample hold by a robot. Kodak DCS Pro 14N has been used in this system to capture 12-bit RGB images with a resolution 4500×3000, but for more glossy material samples capturing HDR images could be required. A practical problem here is the capture time of over 6500 images, which must be multi-fold increased when a multi-exposure technique is used. This problem could be alleviated, when an HDR camera

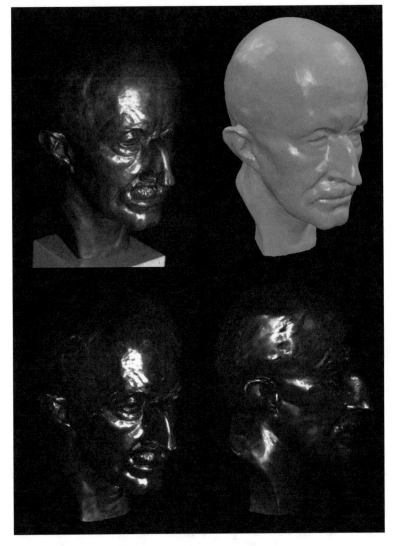

FIGURE 9.18: Digitalization of the Max Planck bust using the acquisition setup [154] shown in Fig. 9.17: (upper left) photograph, (upper right) acquired 3D geometric model, (lower left) rendered image based on the acquired geometry model and spatially varying BRDF distribution for the same viewpoint as the photograph in (upper left), and (lower right) rendered image based on the same acquired model as in (lower right), but illuminated by different lighting and seen from a different viewpoint. Images courtesy of Hendrik P. A. Lensch, Jan Kautz, Michael Goesele, and Hans-Peter Seidel of MPI Informatik and Wolfgang Heidrich of the University of British Columbia. © 2003 ACM, Inc. Used by permission.

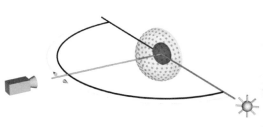

FIGURE 9.19: Acquisition setup for bi-directional texture function capturing [155]. A planar 10 cm×10 cm texture sample is attached to the robot's sampleholder which may change its orientation with respect to a fixed HMI bulb and a rail-mounted CCD camera (Kodak DCS Pro 14N). Images courtesy of Gero Müller, Jan Meseth, Mirko Sattler, Ralf Sarlette, and Reinhard Klein of the University of Bonn.

would be used, which for this particular application should offer very high resolution as well. For more information on BTF acquisition and rendering refer to an excellent survey on this topic by Müeller et al. [155].

Reflectance Field Acquisition

The reflectance field as introduced by Debevec et al. [156] is an 8D function, which relates incoming lighting from the direction (θ_i, ϕ_i) at any point (u_i, v_i) at the surface A to outgoing lighting in the direction (θ_o, ϕ_o) at any point (u_o, v_o) at A. The reflectance field dimensionality can be reduced to 6D by assuming that lighting is distant (a similar assumption as for the environment map lighting in Section 9.2.1), which effectively means that incoming lighting does not vary over the surface of A for each point (u_i, v_i). By making another simplifying assumption that the camera viewpoint is fixed, only a single outgoing lighting direction (θ_o, ϕ_o) is considered for each point (u_o, v_o) at A, what further reduces the reflectance field dimensionality to more tractable 4D. Note that even such a 4D slice over the general 8D reflectance field still provides information on important aspects of light transfer within the material including subsurface scattering. In cinematography and game applications, the human skin is an important example of material, which without modeling of the sub-surface scattering effect has an unnatural plastic look. Debevec et al. [156] demonstrated that the 4D reflectance field of a human face can be reconstructed from a set of images with light rotating around the face at various heights. Essentially a set of basis images for various light directions has been created, which then by their linear combination with different weights enables to render the image of face under

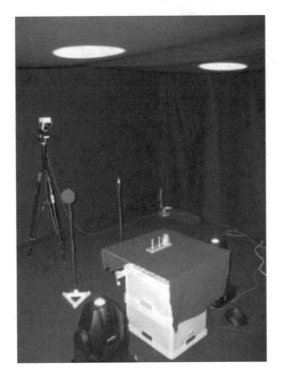

FIGURE 9.20: Photograph of a setup used for capturing of the 4D reflectance field [160]. Spotlight projectors placed on the floor illuminate a tent made of black cloth and indirectly illuminate the captured scene, which is arranged on top of the boxes. A camera mounted on the tripod records the HDR sequences with dynamically changing lighting due to computer-controlled changes in the orientation of the spotlight projectors. Image courtesy of Martin Fuchs, Volker Blanz, Hendrik P. A. Lensch, and Hans-Peter Seidel of MPI Informatik. © 2007 ACM, Inc. Used by permission.

arbitrary lighting and the sub-surface scattering effect is properly considered. Since during the acquisition the human face must remain static it is desirable to use high speed camera. The follow-up research has been focused on lifting the restriction of dimensionality for the captured reflectance fields by allowing arbitrary camera position [157], spatially varying lighting [158], or even full 8D reflectance field [159]. Fuchs et al. [160] showed how to reduce the number of basis images and still achieve the good quality in the scene re-lighting for 4D reflectance fields (with the fixed camera and distant lighting assumptions as in [156]). HDR image capture has been commonly used as it is required to handle strongly glossy objects and improves the overall acquisition accuracy. Figure 9.20 shows the acquisition setup from [160] in which a multi-exposure technique [111] is employed to capture HDR sequences using Jenoptik CEcool or C14plus cameras (refer to Section 3.2 for more details on the C14plus camera). Figure 9.21(left)

FIGURE 9.21:   Rendering of the bottle containing a colored liquid (right) as re-lighted by a real-world environment map (left) [160]. The image has been reconstructed using 1024 HDR images captured for different lighting conditions using the setup from Fig. 9.20. Images courtesy of Martin Fuchs, Volker Blanz, Hendrik P. A. Lensch, and Hans-Peter Seidel of MPI Informatik. © 2007 ACM, Inc. Used by permission.

shows an environment map used to relight the scene with the bottle containing a colored liquid Fig. 9.21(right). Notice subtle light transport effects including anisotropy in the reflectance field due to the interplay of cylindrical bottle's shape with glossy surface material.

Translucent Objects Acquisition

Another important category of real-world materials is translucent objects characterized by complex light scattering inside the material. This multiple light scattering enables us to see light shining through the object and washes out visible surface details by reducing contrast of reflected light. The latter effect is similar to the ambient term in simple reflectance models, but sub-surface scattering may add significant spatially varying and usually low spatial frequency lighting component. Apart from the human skin other examples of translucent materials include milk, marble, and many organic objects such as some fruits. Jensen et al. [162] were the first to address the problem of physics-based translucency modeling and rendering. They proposed an approximation to a diffusion model suitable for rendering of homogeneous materials, and they measured physical parameters required by this model. In their measurement setup, they illuminate material with strong narrow beam of light and capture HDR images using a multi-exposure technique, which is necessary to capture the exponential fall off of scattered light intensity away from the point of illumination (they reported up to five orders of magnitude in the measured light fall-off). Goesele et al. [161] have proposed a measurement setup to capture inhomogeneous translucent object appearance (refer to Fig. 9.22). In their system, they use a narrow laser beam to sequentially illuminate a dense set of locations on the object surface, and the resulting scattered light distribution is captured using an HDR video camera (refer to Fig. 9.23 to see captured sample images for various objects). The use of HDR video camera is mandatory in this application given the amount of images to be captured as well as extremely high dynamic range in scattered lighting. The authors used a Silicon Vision Lars III HDR

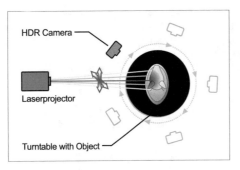

FIGURE 9.22: Acquisition setup for measuring the appearance of inhomogeneous translucent materials [161]. A narrow laser beam, deflected by a high precision 2D galvanometer scanner, sweeps over the object's surface with a sample spacing of about 1 mm. The distribution of scattered lighting for each laser illumination sample is captured by an HDR video camera. For a given camera position all sides of the object are captured using a turntable. This process is repeated for manually changed camera positions, so that the full 360° range of relative laser and camera positions is covered. The two spotlights visible on both sides of the HDR video camera in the right image are used only for object geometry acquisition, which is not discussed here. Images courtesy of Michael Goesele, Hendrik P. A. Lensch, Jochen Lang, Christian Fuchs, and Hans-Peter Seidel of MPI Informatik. © 2004 ACM, Inc. Used by permission.

video camera of resolution 768 × 496 equipped with a high-quality lens to reduce flare effects (refer to Section 3.2 for more details on this camera). The captured data are re-sampled over the vertices of dense mesh, which describes the object geometry, and are used to compute scattered and reflected lighting under arbitrary illumination.

9.3   CONCLUSIONS

In this chapter, we have discussed cross-correlations between developments in computer graphics and HDRI. Realistic graphics and more recently the movie industry relying on digital technology are rich sources of high quality HDR content. In coming years, the role of modern GPUs and game consoles will be increasing in on-line HDR content generation, which will be even more important with the improving availability of HDR display devices (refer to Chapter 7). HDRI contributes to graphics as well by providing captured lighting and object appearance. HDRI-based lighting dominates now in special effects, mixed reality applications, and car advertisement due to much better visual quality of resulting images, good match of virtual and real part of scenography as well as freedom concerning the place and time of HDR light capturing. It can envisioned that soon virtual TV studios, driving simulators, and games will benefit to a greater extent from this technology. In all these applications, the role of HDR video will be increasing since the dynamic aspect of lighting is important in many discussed applications. In

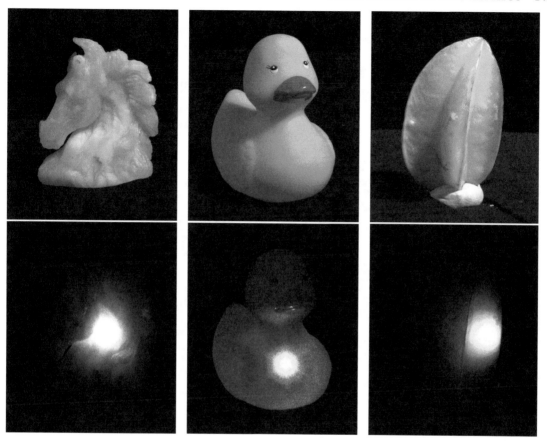

FIGURE 9.23: Translucent Objects Acquisition: (top row) the test objects used to capture their translucent appearance under indoor illumination and (bottom row) the same objects illuminated by a spot-shaped laser beam as captured by an HDR video camera using the acquisition setup shown in Fig. 9.22 [161]. Images courtesy of Michael Goesele, Hendrik P. A. Lensch, Jochen Lang, Christian Fuchs, and Hans-Peter Seidel of MPI Informatik. © 2004 ACM, Inc. Used by permission.

surface reflectance or even more general reflectance field capturing HDRI becomes a standard practice. Here the use of HDR video camera can lead to shortening of the acquisition time which is in particular important when humans or animals are captured. In acquisition setups that require higher sample density such as bi-directional texture function HDR still cameras could be a better choice because at least at present they provide higher image resolution at much lower costs.

CHAPTER 10

Software

To facilitate the work with HDR images and video, Mantiuk et al. [33] have developed a set of software tools that provide a wide range of image and video processing functionality. The tools share a common design pattern based on system pipes which permits to combine them in the form of filters in a processing pipeline, similar to the *netpbm* toolkit. Such a pipeline starts with an input program that reads a list of images and forwards the data in a uniform manner to the next tool. The subsequent tools can perform certain image-processing operations including cropping, rotating, and tone mapping. The last tool in the pipeline usually stores the processed content.

The communication in the pipeline is facilitated by a generic protocol *pfs* whose implementation is offered as a C++ library. The protocol is also straightforward to implement in other languages. The tools exchange data using the pipes commonly supported by many operating systems. Such a design eases the implementation of new tools and permits to transparently combine programs written in various programming languages including MATLAB® and GNU Octave scripts, Perl, Python, and many others. The design principles, including the choice of data representation in the pipeline, are described in more detail in [33].

The main package of the software is *pfstools* and it is currently extended with *pfstmo*, *pfscalibration*, and *HDR Visible Differences Predictor (VDP)*. The whole software is Open Source and can be compiled on several operating systems. It is supported by an active news-group that gathers users and developers.

10.1 PFSTOOLS

pfstools is the main package of the software. It implements the generic communication protocol in the stand-alone library *libpfs*, and contains numerous basic image-processing tools including an HDR capable viewer. *pfstools* supports many HDR and standard file formats including: Radiance RGBE, OpenEXR, Tiff, LogLuv, PFM, PPM, RAW formats of digital cameras, and practically all 8-bit formats through ImageMagick®.

Project page:
`http://www.mpi-inf.mpg.de/resources/pfstools/`

10.2 PFSCALIBRATION

pfscalibration package provides an implementation of the method developed by Robertson et al. [28] for the recovery of the response curve of arbitrary cameras. Tools provided in this package can be used for photometric calibration of both off-the-shelf digital cameras and HDR cameras as described in Section 3.2, and for the recovery of high dynamic range images from the set of low dynamic range exposures as explained in Section 3.1.1.

Project page:
http://www.mpi-inf.mpg.de/resources/hdr/calibration/pfs.html

10.3 PFSTMO

pfstmo package contains implementations of the state-of-the-art tone-mapping operators, including those described in Section 6.1. The implementations are suitable for convenient processing of both static images and animations.

Project page:
http://www.mpi-inf.mpg.de/resources/tmo/

10.4 HDR VISIBLE DIFFERENCES PREDICTOR

HDR Visible Differences Predictor (VDP) belongs to the category of visual metrics, which can predict whether differences between two images are visible to the human observer or not (refer to Chapter 4). Such metrics are used for testing either visibility of information (whether we can see important visual information) or visibility of noise (to make sure we do not see any distortions in images, e.g., due to lossy compression). The unique feature of the HDR VDP is that it can work with the full range of luminance that can be seen by the human eye in the real-world scenes, which effectively means that visual differences between any pair of HDR images can be predicted.

Project page:
http://www.mpi-inf.mpg.de/resources/hdr/vdp

Bibliography

[1] J. Morovic and M. R. Luo. The fundamentals of gamut mapping: A survey. *J. Imaging Sci. Techno.*, Vol. 45(3), pp. 283–290, 2001.

[2] H. Seetzen, W. Heidrich, W. Stuerzlinger, G. Ward, L. Whitehead, M. Trentacoste, A. Ghosh, and A. Vorozcovs. High dynamic range display systems. *ACM Trans. Graph. (Proc. SIGGRAPH)*, 23(3):757–765, 2004.

[3] J. McCann, "Perceptual rendering of HDR in painting and photography," in *Human Vision and Electronic Imaging XIII*, Vol. 6806, SPIE, 2005. Article 30.

[4] R. W. G. Hunt. *The Reproduction of Colour in Photography, Printing and Television: 5th Edition.* Fountain Press, 1995.

[5] G. Wyszecki and W. S. Stiles. *Color Science*, New York: Willey, 2000.

[6] E. Reinhard, G. Ward, S. Pattanaik, and P. Debevec. *High Dynamic Range Imaging: Acquisition, Display, and Image-Based Lighting.* Morgan Kaufmann, 2005.

[7] CIE. *Colorimetry*, Vol. CIE 15.2. International Commision on Illumination, 1986.

[8] M. D. Fairchild. *Color Appearance Models*. Reading, MA: Addison-Wesley, 1998. ISBN 0-201-63464-3.

[9] A. Stockman and L. T. Sharpe. Spectral sensitivities of the middle- and long-wavelength sensitive cones derived from measurements in observers of known genotype. *Visi. Res.*, 40:1711–1737, 2000. doi:10.1016/S0042-6989(00)00021-3

[10] CIE. *A Colour Appearance Model for Colour Management Systems: CIECAM02*, Vol. CIE 159:2004. International Commision on Illumination, 2002.

[11] S. Mann and R. W. Picard, "On being 'undigital' with digital cameras: extending dynamic range by combining differently exposed pictures," in *IS&T's 48th Annual Conf.*, Washington D.C., May 1995. Society for Imaging Science and Technology, pp. 422–428.

[12] P. E. Debevec and J. Malik, "Recovering high dynamic range radiance maps from photographs," in *Proc. SIGGRAPH 97*, Computer Graphics Proceedings, Annual Conf. Ser., 1997, pp. 369–378. doi:10.1145/258734.258884

[13] T. Mitsunaga and S. K. Nayar, "Radiometric self calibration," in *Proc. of IEEE Conf. on Computer Vision and Pattern Recognition*, 1999, pp. 374–380.

[14] G. Krawczyk, K. Myszkowski, and H.-P. Seidel. Lightness perception in tone reproduction for High Dynamic Range images. *Comput. Graph. Forum (Proc. EUROGRAPHICS)*, 24(3):635–645, 2005. doi:10.1111/j.1467-8659.2005.00888.x

[15] S. B. Kang, M. Uyttendaele, S. Winder, and R. Szeliski, High dynamic range video. *ACM Transactions on Graphics (Proc. SIGGRAPH)*, 22(3):319–325, 2003. doi:10.1145/882262.882270

[16] J. Unger and S. Gustavson, "High dynamic range video for photometric measurement of illumination," in *Human Vision and Electronic Imaging XII*, SPIE, Vol. 6501. SPIE, 2007.

[17] S. K. Nayar and T. Mitsunaga, "High dynamic range imaging: spatially varying pixel exposures," in *Proc. IEEE Conf. on Computer Vision and Pattern Recognition*, 2000.

[18] S.K. Nayar and V. Branzoi, "Adaptive dynamic range imaging: Optical control of pixel exposures over space and time," in *Proc. IEEE Int. Conf. on Computer Vision (ICCV 2003)*, 2003, pp. 1168–1175.

[19] S.K. Nayar, V. Branzoi, and T.E. Boult, "Programmable imaging using a digital micromirror array," in *CVPR04*, 2004, pp. I: 436–443.

[20] R. Street, "High dynamic range segmented pixel sensor array," Technical Rep., U.S. Patent 5,789,737, August 1998.

[21] M. Aggarwal and N. Ahuja. Split aperture imaging for high dynamic range. *Proc. Int. Conf. Comput. Vis. (ICCV)*, 2:10–17, 2001.

[22] M. Aggarwal and N. Ahuja. Split aperture imaging for high dynamic range. *Int. J. Comput. Vis.*, 58(1):7–17, 2004. doi:10.1023/B:VISI.0000016144.56397.1a

[23] J. R. Janesick, *Scientific Charge-Coupled Devices*. SPIE, 2001.

[24] R. Ginosar and A. Gnusin. A wide dynamic range CMOS image sensor. *IEEE Workshop on CCD and Advanced Image Sensors*, June 1997.

[25] V. Brajovic, R. Miyagawa, and T. Kanade. Temporal photoreception for adaptive dynamic range image sensing and encoding. *Neural Networks*, 11(7-8):1149–1158, October 1998. doi:10.1016/S0893-6080(98)00070-7

[26] T. Lulé, H. Keller, M. Wagner, and M. Böhm, "LARS II—a high dynamic range image sensor with *a-Si:H* photo conversion layer," in *IEEE Workshop on Charge-Coupled Devices and Advanced Image Sensors*, Nagano, Japan, 1999.

[27] B. Hoefflinger, Editor. *High-Dynamic-Range (HDR) Vision*, (Vol. 26 of *Springer Series in Advanced Microelectronics*), Berlin: Springer, 2007.

[28] M. A. Robertson, S. Borman, and R. L. Stevenson. Estimation-theoretic approach to dynamic range enhancement using multiple exposures. *J. Electron. Imaging*, 12(2):219–228, April 2003. doi:10.1117/1.1557695

[29] G. Krawczyk, M. Goesele, and H.-P. Seidel, "Photometric calibration of high dynamic range cameras," Research Report MPI-I-2005-4-005, Max-Planck-Institut für Informatik, Stuhlsatzenhausweg 85, 66123 Saarbrücken, Germany, April 2005.

[30] M. D. Grossberg and S. K. Nayar, "High dynamic range from multiple images: Which exposures to combine?" in *Proc. ICCV Workshop on Color and Photometric Methods in Computer Vision (CPMCV)*, 2003.

[31] U. Seger, H.-G. Graf, and M. E. Landgraf. Vision assistance in scenes with extreme contrast. *IEEE Micro*, 12(1):50–56, 1993. doi:10.1109/40.210524

[32] M. D. Grossberg and S. K. Nayar. Determining the camera response from images: What is knowable? *IEEE Trans. Pattern Anal. Machine Intell.*, 25(11):1455–1467, November 2003. doi:10.1109/TPAMI.2003.1240119

[33] R. Mantiuk, G. Krawczyk, R. Mantiuk, and H.-P. Seidel, "High dynamic range imaging pipeline: Perception-motivated representation of visual content," in *Human Vision and Electronic Imaging XII*, Vol. 6492 of *Proceedings of SPIE*, B. E. Rogowitz, T. N. Pappas, and S. J. Daly, Eds., SPIE, 2007.

[34] Z. Wang, A. C. Bovik, H. R. Sheikh, and E. P. Simoncelli. Image quality assessment: from error visibility to structural similarity. *IEEE Trans. Image Process.*, 13(4):600–612, 2004. doi:10.1109/TIP.2003.819861

[35] S. Daly, "14: The visible differences predictor: An algorithm for the assessment of image fidelity," in *Digital Images and Human Vision*, A. B. Watson, Ed., MIT Press, 1993, pp. 179–206. ISBN: 0-262-23171-9.

[36] J. Lubin, "A visual discrimination model for imaging system design and evaluation," in *Vision Models for Target Detection and Recognition*, E. Peli, Ed., World Scientific Publishing Company, Inc., 1995, pp. 245–283.

[37] D. M. Chandler and S. S. Hemami. VSNR: A wavelet-based visual signal-to-noise ratio for natural images. *IEEE Trans. IMAGE Process.*, 16(9):2284, 2007. doi:10.1109/TIP.2007.901820

[38] H.R. Sheikh, Z. Wang, L. Cormack, and A.C. Bovik. Live image quality assessment database Release 2. http://live.ece.utexas.edu/research/quality.

[39] R.J. Deeley, N. Drasdo, and W. N. Charman. A simple parametric model of the human ocular modulation transfer function. *Ophthalmol. Physiol. Opt.*, 11:91–93, 1991.

[40] C. Poynton. *Digital Video and HDTV: Algorithms and Interfaces*. Morgan Kaufmann, 2003.

[41] G. Krawczyk, K. Myszkowski, and H.-P. Seidel, "Perceptual effects in real-time tone mapping," in *SCCG '05: Proc. of the 21st Spring Conf. on Computer Graphics*, 2005, pp. 195–202. doi:10.1145/1090122.1090154

[42] G. W. Larson. LogLuv encoding for full-gamut, high-dynamic range images. *J. Graph. Tools*, 3(1):815–30, 1998.

[43] G. McTaggart, C. Green, and J. Mitchell, "High dynamic range rendering in valve's source engine," in *ACM SIGGRAPH 2006 Courses*, 2006, p. 76.

[44] J. Munkberg, P. Clarberg, J. Hasselgren, and T. Akenine-Möller. High dynamic range texture compression for graphics hardware. *ACM Trans. on Graph. (Proc. SIGGRAPH)*, 25(3):698–706, 2006. doi:10.1145/1141911.1141944

[45] C. A. Poynton. *A Technical Introduction to Digital Video*. New York: Wiley, 1996.

[46] R. Mantiuk, K. Myszkowski, and H.-P. Seidel. "Lossy compression of high dynamic range images and video," in *Proc. Human Vision and Electronic Imaging XI*, Vol. 6057 of *Proc. of SPIE*, San Jose, USA, February SPIE, 2006, 60570V.

[47] CIE. *An Analytical Model for Describing the Influence of Lighting Parameters Upon Visual Performance*, Vol. 1. Technical Foundations, CIE 19/2.1. International Organization for Standardization, 1981.

[48] IEC 61966-2-1:1999. *Multimedia Systems and Equipment – Colour Measurement and Management – Part 2-1: Colour Management – Default RGB Colour Space – sRGB*. International Electrotechnical Commission, 1999.

[49] G. Ward. Real pixels. In *Graphics Gems II*, J. Arvo, Ed., Academic Press, 1991, pp. 80–83.

[50] R. Bogart, F. Kainz, and D. Hess, "OpenEXR image file format," in *ACM SIGGRAPH 2003, Sketches & Applications*, 2003.

[51] F. Kainz. Using OpenEXR and the color transform language in digital motion picture production. Technical report, Academy Of Motion Picture Arts and Sciences, 2007.

[52] G. J. Sullivan, H. Yu, S. Sekiguchi, H. Sun, T. Wedi, S. Wittmann, Y. Lee, A. Segall, and T. Suzuki, "New standardized extensions of MPEG4-AVC/H. 264 for professional-quality video applications," in *Proc. ICIP'07*, 2007.

[53] R. Mantiuk, G. Krawczyk, K. Myszkowski, and H.-P. Seidel. Perception-motivated high dynamic range video encoding. *ACM Trans. Graph. (Proc. SIGGRAPH)*, 23(3):730–738, 2004.

[54] ISO-IEC 14496-2. *Information Technology: Coding of Audio-Visual Objects, Part 2: Visual*. International Organization for Standardization, Geneva, Switzerland, 1999.

[55] ISO/IEC 14496-10. *Information Technology: Coding of Audio-Visual Objects, Part 10: Advanced Video Coding*, International Organization for Standardization, Geneva, Switzerland, 2005.

[56] K. E. Spaulding, G. J. Woolfe, and R. L. Joshi. "Using a residual image to extend the color gamut and dynamic range of an sRGB image," in *Proc. IS&T PICS Conference*, 2003, pp. 307–314.

[57] G. Ward and M. Simmons, "Subband encoding of high dynamic range imagery," in *APGV '04: 1st Symposium on Applied Perception in Graphics and Visualization*, 2004, pp. 83–90. doi:10.1145/1012551.1012566

[58] G. Ward and M. Simmons, "JPEG-HDR: A backwards-compatible, high dynamic range extension to JPEG," in *Proc. 13th Color Imaging Conf.*, 2005, pp. 283–290.

[59] Y. Li, L. Sharan, and E. H. Adelson. Compressing and companding high dynamic range images with subband architectures. *ACM Trans. Graph. (Proc. SIGGRAPH)*, 24(3):836–844, 2005. doi:10.1145/1073204.1073271

[60] Z. Wang and A. C. Bovik. A universal image quality index. *IEEE Signal Process Lett.*, 9(3):81–84, March 2002. doi:10.1109/97.995823

[61] A. Segall,"Scalable coding of high dynamic range video," in *Proc. ICIP'07*, 2007.

[62] K. I. Iourcha, K. S. Nayak, and Z. Hong, "System and method for fixed-rate block-based image compression with inferred pixel values," U.S. Patent 5,956,431, 1999.

[63] K. Roimela, T. Aarnio, and J. Itäranta. High dynamic range texture compression. *ACM Trans. Graph. (Proc. SIGGRAPH)*, 25(3):707–712, 2006. doi:10.1145/1141911.1141945

[64] F. Drago, K. Myszkowski, T. Annen, and N. Chiba. Adaptive logarithmic mapping for displaying high contrast scenes. *Comput. Graph. Forum (Proc. of EUROGRAPHICS)*, 24(3):419–426, 2003. doi:10.1111/1467-8659.00689

[65] E. Reinhard, M. Stark, P. Shirley, and J. Ferwerda. Photographic tone reproduction for digital images. *ACM Trans. Graph. (Proc. SIGGRAPH)*, 21(3):267–276, 2002.

[66] E. Reinhard and K. Devlin. Dynamic range reduction inspired by photoreceptor physiology. *IEEE Trans. on Visual. Comput. Graph.*, 11(1):13–24, 2005. doi:10.1109/TVCG.2005.9

[67] G. Ward, H. Rushmeier, and C. Piatko. A visibility matching tone reproduction operator for high dynamic range scenes. *IEEE Trans. Visual. Comput. Graph.*, 3(4):291–306, 1997. doi:10.1109/2945.646233

[68] F. Durand and J. Dorsey. Fast bilateral filtering for the display of high-dynamic-range images. *ACM Trans. Graph. (Proc. SIGGRAPH)*, 21(3):257–266, 2002.

[69] C. Tomasi and R. Manduchi, "Bilateral filtering of gray and colored images." in *Proc. IEEE International Conf. on Computer Vision*, 1998, pp. 836–846.

[70] J. Chen, S. Paris, and F. Durand. Real-time edge-aware image processing with the bilateral grid. *ACM Trans. Graph. (Proc. SIGGRAPH)*, 2007. Article 103.

[71] A. Gilchrist, C. Kossyfidis, F. Bonato, T. Agostini, J. Cataliotti, X. Li, B. Spehar, V. Annan, and E. Economou. An anchoring theory of lightness perception. *Psychol. Rev.*, 106(4):795–834, 1999. doi:10.1037/0033-295X.106.4.795

[72] D. Lischinski, Z. Farbman, M. Uyttendaele, and R. Szeliski. Interactive local adjustment of tonal values. *ACM Trans. Graph. (Proc. SIGGRAPH)*, 25(3):646–653, 2006. doi:10.1145/1141911.1141936

[73] G. Krawczyk, R. Mantiuk, D. Zdrojewska, and H.-P. Seidel, "Brightness adjustment for HDR and tone mapped images," in *The 15th Pacific Conf. on Computer Graphics and Applications*, Computer Graphics and Applications. IEEE, 2007.

[74] R. Fattal, D. Lischinski, and M. Werman. Gradient domain high dynamic range compression. *ACM Trans. Graph. (Proc. SIGGRAPH)*, 21(3):249–256, 2002.

[75] R. Mantiuk, K. Myszkowski, and H.-P. Seidel. A perceptual framework for contrast processing of high dynamic range images. *ACM Trans. Appli. Percept.*, 3:286–308, 2006. doi:10.1145/1166087.1166095

[76] F. Drago, W. L. Martens, K. Myszkowski, and H.-P. Seidel, "Perceptual evaluation of tone mapping operators with regard to similarity and preference," Technical Report MPI-I-2002-4-002, Max-Planck-Institut für Informatik, Im Stadtwald 66123 Saarbrücken, Germany, October 2002.

[77] J. Kuang, H. Yamaguchi, G. M. Johnson, and M. D. Fairchild, "Testing HDR image rendering algorithms," in *Proc. of IS&T/SID 12th Color Imaging Conf.*, 2004, pp. 315–320. doi:10.1145/1015706.1015749

[78] P. Ledda, A. Chalmers, T. Troscianko, and H. Seetzen. Evaluation of tone mapping operators using a high dynamic range display. *ACM Trans. Graph. (Proc. SIGGRAPH)*, 24(3):640–648, 2005. doi:10.1145/1073204.1073242

[79] A. Yoshida, V. Blanz, K. Myszkowski, and H.-P. Seidel, "Perceptual evaluation of tone mapping operators with real-world scenes," in *Human Vision and Electronic Imaging X*, SPIE, Vol. 5666, 2005, pp. 192–203, SPIE.

[80] P. B. Delahunt, X. Zhang, and D. H. Brainard. Perceptual image quality: Effects of tone characteristics. *J. of Electron. Imaging*, 14(2):1–12, 2005.

[81] A. Yoshida, R. Mantiuk, K. Myszkowski, and H.-P. Seidel. Analysis of reproducing real-world appearance on displays of varying dynamic range. *Comput. Graph. Forum (Proc. of EUROGRAPHICS)*, 25(3):415–426, 2006. doi:10.1111/j.1467-8659.2006.00961.x

[82] H. Seetzen, H. Li, L. Ye, W. Heidrich, L. Whitehead, and G. Ward, "25.3: Observations of luminance, contrast and amplitude resolution of displays," in *SID 06 Digest*, 2006, pp. 1229–1233. doi:10.1889/1.2433199

[83] K. Smith, G. Krawczyk, K. Myszkowski, and H.-P. Seidel. Beyond tone mapping: Enhanced depiction of tone mapped HDR images. *Comput. Graph. Forum (Proc. of EUROGRAPHICS)*, 25(3):427–438, 2006. doi:10.1111/j.1467-8659.2006.00962.x

[84] G. Krawczyk, K. Myszkowski, and H.-P. Seidel. Contrast restoration by adaptive countershading. *Comput. Graph. Forum (Proc. of EUROGRAPHICS)*, 26(3):581–590, 2007. doi:10.1111/j.1467-8659.2007.01081.x

[85] J. A. Ferwerda, S. Pattanaik, P. Shirley, and D. P. Greenberg, "A model of visual adaptation for realistic image synthesis," in *Proc. SIGGRAPH 96*, Computer Graphics Proceedings, Annual Conf. Ser., 1996, pp. 249–258. ACM SIGGRAPH/ACM Press. doi:10.1145/237170.237262

[86] N. Goodnight, R. Wang, C. Woolley, and G. Humphreys, "Interactive time-dependent tone mapping using programmable graphics hardware," in *14th Eurographics Symp. on Rendering*, 2003, pp. 26–37.

[87] F. Durand and J. Dorsey, "Interactive tone mapping," in *11th Eurographics Workshop on Rendering*, 2000, pp. 219–230.

[88] P. Irawan, J. A. Ferwerda, and S. R. Marschner, "Perceptually based tone mapping of high dynamic range image streams," in *16th Eurographics Symposium on Rendering*, 2005, pp. 231–242.

[89] G. Damberg, H. Seetzen, G. Ward, W. Heidrich, and L. Whitehead, "3.2: High dynamic range projection systems," in *SID 07 Digest*, 2007.

[90] C. Deter and W. Biehlig, "Scanning laser projection display and the possibilities of an extended color space," in *CGIV 2004 – 2nd European Conf. on Color in Graphics, Imaging and Vision*, pp. 531–535. Berlin: Springer-Verlag, 2004.

[91] J. Agostinelli, M. W. Kowarz, D. Stauffer, T. Madden, , and J. G. Phalen, "Gems: A simple light modulator for high-performance laser projection display," in *Proc. of ITE/SID 13th International Display Workshops (IDW06)*, 2006, pp. 1579–1582.

[92] T.-H. Kim, J. Ahn, and M. G. Choi. Image dequantization: Restoration of quantized colors. *Comput. Graph. Forum (Proc. of EUROGRAPHICS)*, 26(3):619–626, 2007. doi:10.1111/j.1467-8659.2007.01085.x

[93] S. Daly and X. Feng, "Decontouring: Prevention and removal of false contour artifacts," in *Proc. Human Vision and Electronic Imaging IX*, SPIE, Vol. 5292, 2004, pp. 130–149.

[94] S. Daly and X. Feng, "Bit-depth extension using spatiotemporal microdither based on models of the equivalent input noise of the visual system," in *Color Imaging VIII: Processing, Hardcopy, and Applications*, SPIE, Vol. 5008, pp. 455–466, 2003.

[95] S. Bhagavathy, J. Llach, and J. fu Zhai, "Mulit-scale probabilisitic dithering for suppressing banding artifacts in digital images," in *IEEE International Conf. on Image Processing (ICIP)*, 2007, pp. IV–397–400.

[96] C. R. Carlson, E. H. Adelson, and C. H. Anderson, "System for coring an image-representing signal," in *US Patent 4,523,230*. United States Patent and Trademark Office, 1985.

[97] A. O. Akyüz, E. Reinhard, R. Fleming, B. E. Riecke, and H. H. Bülthoff. Do HDR displays support ldr content? A psychophysical evaluation. *ACM Trans. Graph. (Proc. SIGGRAPH)*, 26(3), 2007. Article 38.

[98] L. Meylan, S. Daly, and S. Susstrunk, "The reproduction of specular highlights on high dynamic range displays," in *Proc. 14th Color Imaging Conf.*, 2006.

[99] F. Banterle, P. Ledda, K. Debattista, and A. Chalmers, "Inverse tone mapping," in *18th Eurographics Symposium on Rendering*, 2007, pp. 321–326.

[100] A. G. Rempel, M. Trentacoste, H. Seetzen, H. D. Young, W. Heidrich, L. White-head, and G. Ward. Ldr2Hdr: On-the-fly reverse tone mapping of legacy video and photographs. *ACM Trans. Graph. (Proc. SIGGRAPH)*, 26(3), 2007. Article 39.

[101] L. Meylan, "*Tone mapping for high dynamic range images*," Ph.D. thesis, École Polytech-nique Fédéral de Lausanne, 2006.

[102] L. Meylan, S. Daly, and S. Susstrunk, "Tone mapping for high dynamic range displays," in *Human Vision and Electronic Imaging XII*, SPIE, Vol. 6492, 2007.

[103] H. Farid. Blind inverse gamma correction. *IEEE Trans. Image Process.*, pp. 1428–1433, 2001. doi:10.1109/83.951529

[104] S. Lin, J. Gu, S. Yamazaki, and H.-Y. Shum. Radiometric calibration from a single image. *Conf. Comput. Vis. Pattern Recog. (CVPR'04)*, 2:938–945, 2004.

[105] L. Wang, L. Wei, K. Zhou, B. Guo, and H.-Y. Shum, "High dynamic range image hallucination," in *18th Eurographics Symp. on Rendering*, 2007, pp. 321–326.

[106] I. Drori, D. Cohen-Or, and H. Yeshurun. Fragment-based image com-pletion. *ACM Trans. Graph. (Proc. SIGGRAPH)*, 22(3):303–312, July 2003. doi:10.1145/882262.882267

[107] P. Perez, M. Gangnet, and A. Blake. Poisson image editing. *ACM Trans. Graph. (Proc. SIGGRAPH)*, 22(3):313–318, 2003. doi:10.1145/882262.882269

[108] J. McCann and A. Rizzi, "Veiling glare: the dynamic range limit of hdr images," in *Human Vision and Electronic Imaging XII*, SPIE, Vol. 6492, 2007.

[109] J. Starck, E. Pantin, and F. Murtagh. Deconvolution in astronomy: A review. *Pub. Astron. Soc. Pacific*, 114:1051–1069, October 2002. doi:10.1086/342606

[110] E.-V. Talvala, A. Adams, M. Horowitz, and M. Levoy. Veiling glare in high-dynamic-range imaging. *ACM Trans. Graph. (Proc. SIGGRAPH)*, 26(3), 2007. Article 37.

[111] M. A. Robertson, S. Borman, and R. L. Stevenson, "Dynamic range improvement through multiple exposures," in *Proc. 1999 Int. Conf. on Image Processing (ICIP-99)*, Los Alamitos, CA, 1999, pp. 159–163. doi:10.1109/ICIP.1999.817091

[112] G. J. Ward, F. M. Rubinstein, and R. D. Clear, "A ray tracing solution for diffuse interreflection," in *Computer Graphics (Proceedings of SIGGRAPH 88)*, 1988, pp. 85–92.

[113] G. J. Ward, "The RADIANCE lighting simulation and rendering system," in *Proc. of SIGGRAPH 94*, Computer Graphics Proceedings, Annual Conference Series, pages 459–472. ACM SIGGRAPH / ACM Press, 1994. doi:10.1145/192161.192286

[114] C. M. Goral, K. E. Torrance, D. P. Greenberg, and B. Battaile. Modelling the interaction of light between diffuse surfaces. In *Computer Graphics (Proceedings of SIGGRAPH 84)*, 1984, pp. 213–222.

[115] J. T. Kajiya. The rendering equation. In *Computer Graphics (Proceedings of SIGGRAPH 86)*, 1986, pp. 143–150.

[116] T. Nishita and E. Nakamae. Continuous tone representation of three-dimensional objects taking account of shadows and interreflection. In *Computer Graphics (Proceedings of SIGGRAPH 85)*, 1985, pp. 23–30.

[117] R. Siegel and J. R. Howell. *Thermal Radiation Heat Transfer* Washington D.C.: Hemisphere Publishing Corp., 1981.

[118] P. R. Tregenza. The Monte Carlo method in lighting calculations. *Lighting Res. Technol.*, 15(4):163–170, 1983.

[119] D. Stanger. Monte Carlo procedures in lighting design. *J. Illuminating Eng. Soc.*, 13(4):368–371, 1984.

[120] N. J. Miller, P. Y. Ngai, and D. D. Miller. The application of computer graphics in lighting design. *J. Illumin. Eng. Soc.*, 14(1):6–26, 1984.

[121] J. Tumblin and H. E. Rushmeier. Tone reproduction for realistic images. *IEEE Comput. Graph. Appl.*, 13(6):42–48, November 1993. doi:10.1109/38.252554

[122] G. Ward. A contrast-based scalefactor for luminance display. *Graphics Gems IV*, 1994, pp. 415–421.

[123] A. Ghosh, M. Trentacoste, H. Seetzen, and W. Heidrich, "Real illumination from virtual environments," in *16th Eurographics Symp. on Rendering*, pp. 243–252, 2005.

[124] G. Müller, Ralf Sarlette, and Reinhard Klein, "Procedural editing of bidirectional texture functions," in *18th Eurographics Symp. on Rendering*, pp. 219–230, 2007.

[125] R. W. Fleming, R. O. Dror, and E. H. Adelson. Real-world illumination and the perception of surface reflectance properties. *J. Vis.*, 3(5):347–368, 2003. doi:10.1167/3.5.3

[126] E. Land and J. McCann. Lightness and retinex theory. *J. Opt. Soc. Am.*, 61(1):1–11, January 1971.

[127] B. K. P. Horn. Determining lightness from an image. *Comput. Graph. Image Process.*, 3(1):277–299, 1974. doi:10.1016/0146-664X(74)90022-7

[128] J. A. Ferwerda, S. H. Westin, R. C. Smith, and R. Pawlicki, "Effects of rendering on shape perception in automobile design," in *ACM Siggraph Symp. on Applied Perception in Graphics and Visualization 2004*, 2004, pp. 107–114. doi:10.1145/1012551.1012570

[129] SpheronVR. *http://www.spheron.com/*.

[130] P. Vangorp, J. Laurijssen, and P. Dutre. The influence of shape on the perception of material reflectance. *ACM Trans. Graph. (Proc. SIGGRAPH)*, 26(3), 2007. Article 77.

[131]   P. Debevec. Rendering synthetic objects into real scenes: Bridging traditional and image-based graphics with global illumination and high dynamic range photography. In *Proceedings of SIGGRAPH 98*, Computer Graphics Proceedings, Annual Conference Series, 1998, pp. 189–198. doi:10.1145/280814.280864

[132]   G. Ramanarayanan, J. Ferwerda, B. Walter, and K. Bala. Visual equivalence: Towards a new standard for image fidelity. *ACM Trans. Graph. (Proc. SIGGRAPH)*, 26(3), 2007. Article 76.

[133]   R. Ramamoorthi and P. Hanrahan, "An efficient representation for irradiance environment maps," in *Proc. ACM SIGGRAPH 2001*, (Computer Graphics Procs., Annual Conf. Ser., August 2001, pp. 497–500.

[134]   P.-P. Sloan, J. Kautz, and J. Snyder. Precomputed radiance transfer for real-time rendering in dynamic, low-frequency lighting environments. *ACM Trans. Graph. (Proc. SIGGRAPH)*, 21(3):527–536, 2002.

[135]   R. Ng, R. Ramamoorthi, and P. Hanrahan. Triple product wavelet integrals for all-frequency relighting. *ACM Trans. Graph. (Proc. SIGGRAPH)*, 23(3):477–487, 2004.

[136]   P.-P. Sloan, J. Hall, J. Hart, and J. Snyder. Clustered principal components for precomputed radiance transfer. *ACM Trans. Graph. (Proc. SIGGRAPH)*, 22(3):382–391, 2003. doi:10.1145/882262.882281

[137]   Z. Ren, R. Wang, J. Snyder, K. Zhou, X. Liu, B. Sun, P.-P. Sloan, H. Bao, Q. Peng, and B. Guo. Real-time soft shadows in dynamic scenes using spherical harmonic exponentiation. *ACM Trans. Graph. (Proc. SIGGRAPH)*, 25(3):977–986, 2006. doi:10.1145/1141911.1141982

[138]   V. Havran, M. Smyk, G. Krawczyk, K. Myszkowski, and H.-P. Seidel, "Interactive system for dynamic scene lighting using captured video environment maps," in *16th Eurographics Symposium on Rendering*, 2005, pp. 31–42.

[139]   S. Hiller, O. Deussen, and A. Keller, "Tiled blue noise samples," in *Vision, Modeling, and Visualization*, 2001, pp. 265–272.

[140]   T. Kollig and A. Keller, "Efficient illumination by high dynamic range images," in *14th Eurographics Symp. on Rendering*, 2003, pp. 45–51.

[141]   G. S. Fishman. *Monte Carlo: Concepts, Algorithms, and Applications*, Berlin: Springer, 1996.

[142]   M. Pharr and G. Humphreys, "Infinite area light source with importance sampling," in an Internet publication accompanying the book *Physically Based Rendering from Theory to Implementation*, http://pbrt.org/plugins.php, 2004.

[143]   V. Havran, K. Dmitriev, and H.-P. Seidel, "*Goniometric Diagram Mapping for Hemisphere*," in Short Presentations (Eurographics 2003), 2003.

[144] J. Kannala and S. Brandt, "A generic camera calibration method for fish-eye lenses," in *Proc. 2004 Virtual Reality*. IEEE.

[145] S. Laine, H. Saransaari, J. Kontkanen, J. Lehtinen, and T. Aila, "Incremental instant radiosity for real-time indirect illumination," in *18th Eurographics Symposium on Rendering*, 2007, pp. 277–286.

[146] T. Grosch, T. Eble, and S. Mueller, "Consistent interactive augmentation of live camera images with correct near-field illumination," in *ACM Symposium on Virtual Reality Software and Technology (VRST 2007)*. ACM, 2007.

[147] G. Greger, P. S. Shirley, P. M. Hubbard, and D. P. Greenberg. The irradiance volume. *IEEE Computer Graph. Appl.*, 18(2):32–43, 1998. doi:10.1109/38.656788

[148] L. Wan, T.-T. Wong, and C.-S. Leung, "Spherical Q2-tree for sampling dynamic environment sequences," in *16th Eurographics Symposium on Rendering*, pages 21–30, 2005.

[149] M. Korn, M. Stange, A. von Arb, L. Blum, M. Kreil, K.-J. Kunze, J. Anhenn, T. Wallrath, and T. Grosch, "Interactive augmentation of live images using a hdr stereo camera," in Stefan Mueller and Gabriel Zachmann, editors, *Dritter Workshop Virtuelle und Erweiterte Realitaet der GI-Fachgruppe VR/AR, 25–26 September 2006, Koblenz*, 2006, pp. 107–118. Shaker Verlag.

[150] I. Sato, Y. Sato, and K. Ikeuchi. Acquiring a radiance distribution to superimpose virtual objects onto a real scene. *IEEE Trans. Visual. Comput. Graph.*, 5(1):1–12, January-March 1999. doi:10.1109/2945.764865

[151] F. Pellacini, J. A. Ferwerda, and D. P. Greenberg, "Toward a psychophysically-based light reflection model for image synthesis," in *Proc. ACM SIGGRAPH 2000* (Computer Graphics Proceedings, Annual Conf. Ser.), pp. 55–64.

[152] S. R. Marschner, S. H. Westin, E. P. F. Lafortune, K. E. Torrance, and D. P. Greenberg, "Image-based BRDF measurement including human skin," in *9th Eurographics Workshop on Rendering*, 1999, pp. 131–144.

[153] W. Matusik, H. Pfister, M. Brand, and L. McMillan, "Efficient isotropic BRDF measurement," in *14th Eurographics Symposium on Rendering*, 2003, pp. 241–248.

[154] H. P. A. Lensch, J. Kautz, M. Goesele, W. Heidrich, and H.-P. Seidel. Image-based reconstruction of spatial appearance and geometric detail. *ACM Trans. Graph.*, 22(2):234–257, 2003. doi:10.1145/636886.636891

[155] G. Müller, J. Meseth, M. Sattler, R. Sarlette, and R. Klein. Acquisition, synthesis, and rendering of bidirectional texture functions. *Comput. Graph. Forum*, 24(1):83–110, 2005. doi:10.1111/j.1467-8659.2005.00830.x

[156]   P. Debevec, T. Hawkins, C. Tchou, H.-P. Duiker, W. Sarokin, and M. Sagar, "Acquiring the reflectance field of a human face," in *Proc. ACM SIGGRAPH 2000*, (Computer Graphics Procs., Annual Conf. Ser.), 2000, pp. 145–156.

[157]   W. Matusik, H. Pfister, A. Ngan, P. Beardsley, R. Ziegler, and L. McMillan. Image-based 3D photography using opacity hulls. *ACM Trans. Graph. (Proc. SIGGRAPH)*, 21(3):427–437, 2002.

[158]   V. Masselus, P. Peers, P. Dutré, and Y. D. Willems. Relighting with 4D incident light fields. *ACM Trans. Graph. (Proc. SIGGRAPH)*, 22(3):613–620, 2003. doi:10.1145/882262.882315

[159]   G. Garg, E.-V. Talvala, M. Levoy, and H. P. A. Lensch, "Symmetric photography: Exploiting data-sparseness in reflectance fields," in *17th Eurographics Symposium on Rendering*, 2006, pp. 251–262.

[160]   M. Fuchs, V. Blanz, H. P. A. Lensch, and H.-P. Seidel. Adaptive sampling of reflectance fields. *ACM Trans. Graph.*, 26(2), 2007. Article 10. doi:10.1145/1243980.1243984

[161]   M. Goesele, H. P. A. Lensch, J. Lang, C. Fuchs, and H.-P. Seidel. DISCO: acquisition of translucent objects. *ACM Trans. Graph. (Proc. SIGGRAPH)*, 23(3):835–844, 2004. doi:10.1145/1015706.1015807

[162]   H. W. Jensen, S. R. Marschner, M. Levoy, and P. Hanrahan, "A practical model for subsurface light transport," in *Proc. ACM SIGGRAPH 2001*, (Computer Graphics Proc., Annual Conf. Ser.), 2001, pp. 511–518.

Author Biography

Karol Myszkowski is a senior researcher in the Computer Graphics Group of Max-Planck-Institut für Informatik. From 1993 to 2000 he served as a tenured Associate Professor at the University of Aizu, Japan, and from 1985 to 1993 he worked as a Research Associate and then an Assistant Professor at the Szczecin University of Technology, Poland.

In the period 1986-1992 he collaborated with Japanese company Integra, Inc. developing rendering software for such customers as Toshiba Lighting, Shiseido, Matsushita Electric, Kandenko, and others. He received his MSc degree in control engineering from the Szczecin University of Technology in 1983, and his PhD and habilitation degrees in computer science from Warsaw University of Technology (Poland) in 1991 and 2001, respectively. His research interests include perception issues in graphics, high-dynamic range imaging, global illumination, rendering and animation. Myszkowski has written over 80 refereed publications on these subjects and has served on numerous program committees. He was the Program Committee co-chair of Eurographics Rendering Symposium in 2001, ACM Applied Perception in Graphics and Visualization in 2008, and Spring School of Computer Graphics in 2008. Myszkowski has supervised over 30 graduate and undergraduate research projects.

Rafal Mantiuk (PhD from the Max-Planck-Institut for Computer Science, Germany; Msc in Computer Science from the Szczecin University of Technology, Poland) is postdoctoral fellow at the University of British Columbia, Canada. He combines in his research the aspects of human perception, color appearance, image processing, and computer graphics to address the problems of future imaging systems, in which the human eye rather than technology is the major limiting factor. During the last five years he has been involved in research at Max-Planck-Institut for Computer Science, Sharp Laboratories of America, and BrightSide Technologies (Dolby Canada). Rafal authored several patents and papers on high-dynamic range (HDR) image and video compression (Proc. of SIGGRAPH'04 and '06), tone-mapping (ACM TAP, EG'06), developed a fidelity metric for HDR images (HDR-VDP) and co-maintains popular software for HDR processing—pfstools. In 2006 he was granted the Heinz Billing Award for his work.

Grzegorz Krawczyk received PhD from the Max-Planck-Institut for Computer Science, Germany and MSc in Computer Science from the Szczecin University of Technology, Poland. During the last five years he has been involved in the research projects in collaboration with the

University of British Columbia and BrightSide Technologies (Dolby Canada). His research focuses on the insightful application of knowledge about human visual system to assure high fidelity in high dynamic range (HDR) images and video. Grzegorz authored several papers on HDR tone mapping (Eurographics '05, '06, and '07) and HDR video compression (ACM Siggraph '04). He also co-maintains a popular software for HDR processing, capture and tone mapping—pfstools, pfscalibration and pfstmo.

Printed in the United States
by Baker & Taylor Publisher Services